ROGER STEVENSON
JANUARY, 1996

WEATHER

"Every wind has its Weather"

Francis Bacon, 1561–1626

WEATHER

B. W. Atkinson & Alan Gadd

Foreword by
Dr. Frank Field

Weidenfeld & Nicolson
New York

Editor *James Hughes*
Art Editor *Nigel O'Gorman*
Picture Research *Andrea Rittweiler*
Sheila Corr Millicent Trowbridge
Assistant Editor *Julia Gorton*
Assistant Designer *Christopher Howson*
Production *Androulla Pavlou*
Maps and diagrams by
Mulkern Rutherford and Eugene Fleury

Edited and designed by
Mitchell Beazley International Ltd
Artists House, 14-15 Manette St,
London W1V 5LB

WEATHER
A Mitchell Beazley Earth Science Handbook

Published by Weidenfeld & Nicolson, New York
A Division of Wheatland Corporation
10 East 53rd Street
New York, NY 10022

Library of Congress Cataloging-in-Publication Data
Atkinson, Bruce Wilson.
    Weather
    (A Mitchell Beazley earth science handbook)
    Includes index.
    1. Weather–Popular works. 2. Weather
forecasting–Popular works. I. Gadd, A. J.
II. Title. III. Series.
QC981,2,A85 1987   551,6   86-19073

Filmsetting by
Hourds Typographica, Stafford, England
Origination by
Gilchrist Bros. Ltd, Leeds, England
Printed and bound by
Printer Portuguesa LDA, Portugal

First American Edition 1987
10 9 8 7 6 5 4 3 2 1

Foreword

Weather has become of great interest to millions of
Americans who have built homes along ocean and lake
fronts. These residents have learned through experience
that weather conditions may have an enormous impact on
their lives. Especially storms.

Just as many millions have taken up hobbies in which the
daily changing weather patterns are forces to be reckoned
with. For the weekend flier and for the boat owner, an
unforeseen weather event can affect their safety and
well-being.

There are countless others who look down at intricate
cloud patterns outside an airliner window, or gaze at the
sky and marvel at the changing colors and hues of drifting
clouds. Interest in weather has become universal. It's a
subject integral to every news broadcast.

For the past twenty eight years, I have enjoyed
presenting over twenty thousand weather reports, and I
have learned that the hunger for weather knowledge is
never satisfied. The daily mail is proof of that!

- What's the difference between a tornado and a
  hurricane?
- How do you compute the relative humidity?
- What are those symbols on the weather map?
- Which are the clouds on the satellite picture? The dark
  areas or the light ones?
- Why does the weather move from west to east?

Well, here are the answers to those questions and many
more. And you don't have to be a mathematician or
physicist to understand them. The authors Bruce Atkinson
and Alan Gadd are meeting you halfway. For those who
have a curiosity concerning the weather and want to learn
more, this is an ideal book with which to begin.

For those who have a more scientific background and are
serious about learning more about meteorology, there is
much information to be gained here.

Once you sample some of the mysteries of what makes
weather, your knowledge will grow, and you will probably
begin questioning your favorite weatherperson's forecasts.
Who knows? You may turn into the family's weather oracle.

Weather will be a welcome addition to your bookshelf. I
write this, even though it may produce countless
Monday-morning quarterbacks who will call me to task for
any forecasts that turn out to be wrong.

Dr Frank Field

CONTENTS

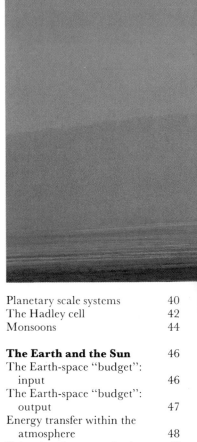

Introduction: World weather patterns

To most of us, weather means sunshine, cloud, wind, heat and cold, rain, frost, fog, snow and hail. These elements impinge directly on us, but the general term of weather includes many other atmospheric characteristics, such as *humidity** and *evaporation*. We experience weather at the Earth's surface, where we live, and for centuries all our observations of the atmosphere were made from the surface. But today meteorologists can observe the weather from below, within and above the atmosphere, and this allows them to describe, analyze and forecast the weather far better than in the past.

Air masses

Over the centuries, much weather lore has evolved, some fanciful, some eminently sensible. A particularly appropriate saying is, "Every wind has its weather." Herein lies a key to weather analysis and forecasting. It tells us that sunshine, warmth, cold, rain and other elements are related to the winds – a far from trivial discovery. Meteorologists find it useful to analyze these important features in terms of the *masses* and the winds themselves.

Air masses are huge volumes of the atmosphere in which horizontal gradients of temperature and humidity are comparatively small. They form over large, homogeneous surfaces such as ice (in Greenland), tundra and forest (in Canada), or the oceans. As a consequence they become relatively cold, dry, humid or warm, as the case may be. Using this simple classification it is possible to divide the atmosphere at any given time into areas dominated by particular air masses. These air masses give particular combinations of weather, such as biting cold, dry air over a continent in winter, or warm, muggy air over coastal areas in summer. *Fronts*, the familiar lines of cloud and rain seen on all weather maps, may form at the confluence of different air masses.

Once weather of a certain character has formed in an air mass, it tends to be carried by the wind and so is exported to other areas. Herein lies the truth of the saying, "Every wind has its weather." This basic

Cold air masses
1 Arctic
2 Polar maritime
3 Polar continental
Cold sea currents
4 Kamchatka
5 California
6 Labrador
Warm air masses
7 Tropical maritime
8 Tropical continental
Warm sea currents
9 North Pacific drift
10 Caribbean
11 Gulf Stream
Hurricanes

Air masses derive their temperature and humidity characteristics from their source areas: cold raw *arctic* air moves into N. Canada; slightly warmer *polar maritime* air sweeps into NW and NE America; *polar continental* air forms over N Canada, causing very cold bursts far to the south; *tropical maritime* air forms over the two oceans, moving inland; *tropical continental* air forms over West USA, giving dry climates.

*Words and terms defined in the glossary occur in *italics* on their first appearance.

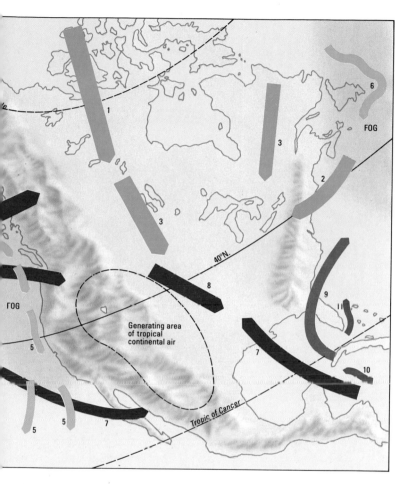

idea applies across a whole range of scales. Tropical air masses originate from low latitudes and polar air masses from high latitudes. Maritime and continental air masses result from the large-scale pattern of land and sea.

Patterns of wind

Related to this broad distribution of air masses is that of the surface winds. Within the tropics, the trade winds blow from latitude 25–30° toward the equator. Between 30° and about 70° the winds are mainly westerly (that is, they blow from the west). So the world's weather exhibits certain gross patterns: hot and wet in equatorial areas (10° latitude either side of the equator); hot and dry between 10° and 35° latitude; cool and wet between 35° and 75° latitude; and predominantly cold and dry poleward of 75°. Superimposed on these simple patterns are many variations, including those due to land-sea distribution, the northern and southern hemisphere, the seasons, and weather systems.

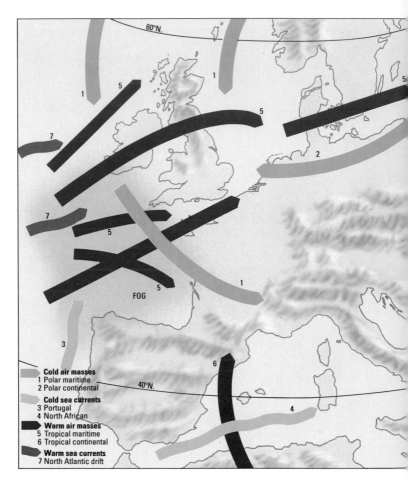

Cold air masses
1 Polar maritime
2 Polar continental

Cold sea currents
3 Portugal
4 North African

Warm air masses
5 Tropical maritime
6 Tropical continental

Warm sea currents
7 North Atlantic drift

The northern hemisphere contains more land than the southern. This results in more clear-cut continental-oceanic contrasts in temperature, humidity and precipitation (an overall term for snow, rain, hail, etc.). The North Pole lies in the Arctic Ocean, which freezes over millions of square miles in winter. The southern hemisphere has virtually no land at all in the critical latitudes 50° to 70°S, and this encourages very strong and persistent westerly winds in that part of the world. In contrast to northern polar areas, Antarctica is not only a massive continent but also has the highest elevation. This has dramatic effects on its temperatures, strongly influencing the wind and weather of the southern hemisphere and possibly the whole globe.

Seasonal weather

The seasons are due to the Earth's movement around the Sun. The Sun appears to us to cross latitudes throughout the year because the Earth's axis is tilted in relation to the plane of its path round

Major air masses over Europe: *polar maritime* air in the NW gives clear, cool days often with good visibility; warmer, still moist *tropical maritime* air is slightly less frequent; *polar continental* air with low winter temperatures forms over mainland Europe, and may replace milder Oceanic conditions with biting cold; further south the Sahara breeds *tropical continental* air that sometimes moves north to give gloriously hot summer weather.

the Sun. Hence in June the midday sun is overhead at latitude $23\frac{1}{2}°$N (Tropic of Cancer) and in December it is overhead at latitude $23\frac{1}{2}°$S (Tropic of Capricorn). It is overhead at the equator twice a year in March and September. This apparent movement of the Sun means that the distribution of winds and weather also moves. In June the patterns lie at their most northward position and in December they lie at their most southward position. It is this movement which causes the weather of any one place to vary from summer to winter. The particular character of any season in any locality results from the third major source of variation mentioned above, the frequency of weather systems of different types.

North America experiences most types of air masses. In winter very cold, dry continental polar air develops over Canada and may sweep southward as far as Florida. On the West Coast, mild, moist tropical maritime air dominates the weather and climate and seasonal contrasts are fairly small. The Gulf States receive very warm, moist air from the ocean, particularly in summer, and this can lead to thunderstorms. The East Coast also receives warm, moist air in summer and often experiences hurricanes spawned over the Atlantic Ocean. In winter, cold northerly air masses moving south frequently result in very heavy snowstorms.

Europe, and particularly the British Isles, also receives a variety of air masses. It has been called the "battleground of the air masses." In winter, raw airstreams from the Arctic or continental Europe may give cold, snowy weather, but more frequently the weather results from maritime air masses from the west – relatively cold if from Greenland and relatively warm if from the Azores. These air masses are also frequent in summer and explain the relative lack of high temperatures in northwest Europe. Occasionally continental tropical air does reach this part of the world, sometimes from as far away as the Sahara desert. At such times, very hot, dry sunny weather occurs, such as in the summer of 1976.

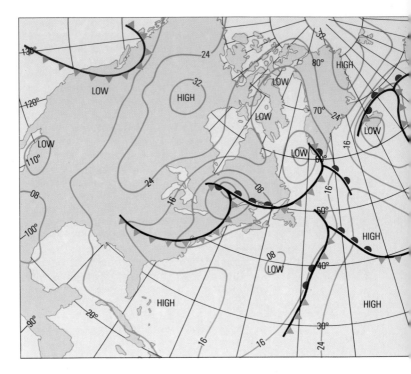

The weather chart

To show the pattern of the weather at a particular time on a particular day, meteorologists use weather charts like the one shown here. The chart summarizes some very important facts about the movement of air near the surface of the Earth and the consequent distribution of warm and cold air, cloud and rain. Similar weather charts are used for higher levels in the atmosphere, but it is the surface chart in particular that has become familiar through newspapers and television.

The continuous lines on a surface weather chart are called *isobars* because they join up points with equal *pressure* – points where the barometer gives the same reading. Pressure is the weight of the air above pressing down on the surface. At land stations a correction is made before the isobars are drawn so that what the chart shows is the pressure at sea level. A choice has to be made about how many lines to draw. On the chart shown here the isobars are drawn at 8 *millibar* intervals, but for working charts 4 millibar intervals are more usual.

On the surface chart, centers of *high* (H) and *low* (L) pressure are marked. In a high (also called an *anticyclone*) the skies are often clear, with dry and settled weather. In a low (also called a *cyclone* or a *depression*) there is likely to be cloud and rain.

The wind near the surface blows approximately along the isobars. If you stand with your back to the wind, low pressure is on the left in the northern hemisphere and on the right in the south-

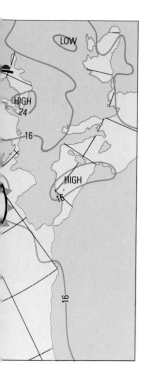

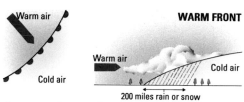

WARM FRONT

As a warm front arrives, cloud gradually thickens ahead of a broad band of rain or snow.

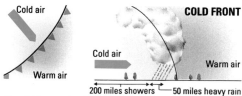

COLD FRONT

A cold front often brings a narrow band of sudden heavy rain, followed by showers in the cold air.

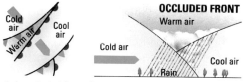

OCCLUDED FRONT

An occluded front forms when a cold front overtakes a warm front and lifts the warm air.

ern hemisphere. Thus the winds blow clockwise around a high and counterclockwise, around a low in the northern hemisphere, with the opposites in the southern hemisphere. The wind direction is actually slightly across the isobars toward low pressure.

The strength of the wind is greatest where the isobars are closest together. Where the isobars are far apart the winds are light, and this is generally the case around high pressure centers. Around a low, very strong winds may occur, although right at the center of the low the wind speed drops almost to nothing.

Fronts

The movement of the air around highs and lows brings cold air masses equatorward and warm air masses poleward. At the confluence of air from different sources, sharp temperature changes may develop and elongated regions of cloud and rain (or snow) are formed. These regions are marked on the weather chart as cold, warm or occluded fronts. At a cold front, cold air is advancing. At a warm front, warm air is advancing. At an *occluded* front, a cold front has caught up on a warm front and the warm air has been lifted away from the surface.

The basic format of isobars, highs and lows, and fronts is internationally agreed. For local use other information is added, using various symbols to denote temperature, wind speed and direction, sunshine, cloud, fog, rain, snow, sleet, hail and thunder. The symbols used vary from country to country.

Weatherwatch: hurricanes

A hurricane is born

The first hint that a storm might be brewing came from a shoebox-size radio transmitter carried by a weather balloon 20 miles into the air over Dakar, Senegal. A wave of low pressure – one of about 60 that originate over western Africa each hurricane season – was rolling off the coast for a long passage across the Atlantic Ocean.

Although that low-pressure wave was no secret to meteorologists, its journey over the warm ocean waters did not take on any special significance until September 16, 1985. That was the date when photographs sent to Earth by the European METEOSAT satellite showed that the wave had developed a low-level circulation – a distinct pattern of *convection* and cyclonic motion around the core of low pressure.

In Miami, Florida, the forecasters at the National Hurricane Center (NHC) knew well that September storms spawned near the Cape Verde Islands off West Africa often grew to become dangerous behemoths, generating winds and storm surges that over the years have killed thousands in the Caribbean and the United States. They studied the satellite pictures, noted the movement of clouds in the easterly trade winds, and decided to mention the disturbance in the day's Tropical Weather Outlook.

Later, that reference would pinpoint the birth of Hurricane Gloria, which for three days teased and terrorized some 26 million people along the populous East Coast of the United States. At one point during its menacing approach to the US coast, a National Oceanic and Atmospheric Administration (NOAA) reconnaissance airplane inside the storm measured its central pressure at 918.6

Hurricane over Mississippi, seen from an orbiting space shuttle. More than 1.5 million people were involved in the most extensive evacuation in US history when Hurricane Elena threatened the coastline from Gulfport to Biloxi, Mississippi, during the 1985 Labor Day weekend. Good collaboration between warning, communications and evacuation systems ensured that no one was killed, although the hurricane claimed four lives elsewhere.

millibars, which translates into winds of about 145mph (233kph). When hurricane specialists at the NHC confirmed that no lower pressure had ever been recorded in that part of the Atlantic, Gloria was dubbed "storm of the century."

In the end, however, Hurricane Gloria seems to have deserved the distinction less for the ferocity of its winds than for its power as a news event. Clearly, it was a dangerous, deadly and destructive hurricane, blamed for eight deaths and up to $1 billion in property damage. It was not, however, the catastrophic killer once feared. (Juan, a minimal hurricane that hit the Gulf Coast of the United States in late October, caused 12 deaths and $1.5 billion in damage.)

The US hurricane watch

In an average hurricane season, which nominally runs from June through to November, more than 100 disturbances with hurricane potential are observed in the Atlantic, Gulf of Mexico and Caribbean Sea. Of these, about 10 reach the tropical storm stage, and only about six mature into hurricanes. A tropical storm earns a name when its sustained winds reach a minimum of 39mph (62kph). It becomes a hurricane when its winds increase to 74mph (119kph).

The responsibility for tracking and warning of tropical storms lies with the six-man team of hurricane specialists of the NHC, a branch of the National Weather Service (NWS). When a tropical depression is identified, the US Air Force and Navy are notified by telephone, and the location of the depression is reported to the world via the High Seas Broadcast, a roundup of marine conditions sent out in Morse code from Washington. News organizations also receive word of the budding storm over the NOAA Weather Wire.

At the tropical storm stage, forecasters call the National Weather Service chiefs in Washington, the NWS regional office in Fort Worth, Texas, and any member nations of the World Meteorological Organization (WMO) that may be affected by the storm within three days. Region 4 members of the WMO include the United States, Mexico, Canada, and several Central and South American countries.

Working in a suburban Miami office building topped by satellite dishes and radar antennas, NHC forecasters sift through a massive flow of data to gauge a storm's intensity and follow its path. Among the tools now routinely used are eye-in-the-sky satellites, land-based radar, aircraft loaded with sensing equipment, a vast network of computer-run data buoys, complex statistical models that make projections based on past storms, and, occasionally, reports from beleaguered ships at sea.

Also important, of course, is the skill of the hurricane specialists, all senior forecasters, who constitute an elite within the ranks of the NWS under the leadership of Neil L. Frank, NHC director since 1974. The hurricane specialists emphasize their role as storm trackers rather than predictors or prognosticators.

The Hurricane Center issues projections on the tracks of storms from the formation of a tropical depression. Every six hours, specialists issue public advisories that include the storm's position, highest winds, and forward motion. If a hurricane watch or warning is posted, the advisories come every three hours. Beginning with a depression, forecasters also issue marine advisories that make projections on the storm's possible position in 12, 24, 48, and 72 hours.

While the introduction in 1984 of a system of landfall probabilities has reduced public pressure on forecasters for clairvoyance, the perplexing question of where a storm will come ashore is one that everyone wants answered. And never was that question asked by more people, with more urgency, than during the reign of Hurricane Gloria.

A fullblown hurricane
Four days west of the Cape Verde Islands, the embryonic storm came within range of the Geostationary Operational Environmental Satellite (GOES), which feeds forecasters with continuous pictures of the Earth from 22,500 miles (36,000km) over the equator. From the swirling movement of clouds, forecasters clearly saw the fledgling storm intensify. Thunderstorms associated with the low had become larger and more numerous, atmospheric pressure was falling, and the circulation of winds around the center had become more organized; The satellite photographs showed that the depression was moving west at a speed of 14 knots.

On September 20, hurricane specialists ordered the US Air Force to investigate the storm. The next day a C-130 turboprop based in Biloxi, Mississippi, located the center of the disturbance about 460 miles (740km) east of the Caribbean's Leeward and Windward Islands. Based on its central pressure and wind speed, the depression was upgraded to a tropical storm, the sixth of the season, and given the name Gloria. A hurricane watch was issued for the northern Leeward Islands. Over the next six days, Air Force and NOAA planes would make 54 penetrations of the storm's center, providing forecasters with a center fix on the average of once every 2.7 hours.

When on the following day, September 22, another reconnaissance flight measured winds of 78mph (125kph), Gloria was pronounced a fullblown hurricane. Warnings were posted for the watch area.

That afternoon, the steering currents guiding Gloria changed its course from due west to west-northwest, a path that spared the Leeward Islands while pointing to the Bahamas and the US East Coast beyond. Up until now, advisories from the NHC had mainly concerned shipping and the residents of far-flung Caribbean islands. Now, however, with its winds growing strong and its path steady, the storm began to attract a much larger audience.

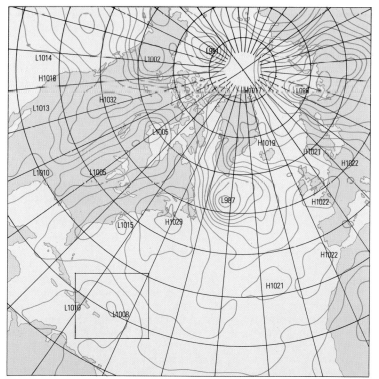

Gloria's low of 1008 millibars is located just north of Puerto Rico on this weather analysis chart, produced soon after noon on September 23 1985. Such a low, at that latitude, over the sea and at that time of the year, invariably signals hurricane conditions.

A forecast plotted
against later analysis
contrasts Gloria's predicted
and actual progress up the
eastern seaboard of the
United States. The plot also
shows how the hurricane
moved more quickly once it
was out of the tropics.

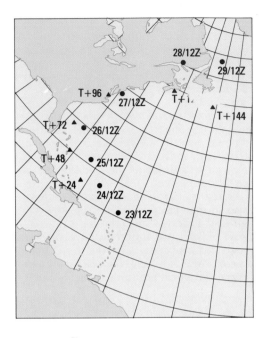

▲ T+96
Forecast positions

● 27/12Z
Observed midday
positions/Date/Time

Hurricane alert

On September 24, Gloria veered north again, skirting the Bahamas. Now only mainland North America lay ahead, and it was too
big to miss. Gloria had tightened up into a dangerous hurricane.
That evening, a NOAA p-3 Orion "flying laboratory" out of Miami
measured the storm's central pressure at a record low. Located
east of the Bahamas, just 600 miles (965km) due east of Miami, the
hurricane's winds were blowing at 145mph (233kph).

Over three days, NHC director Frank gave 276 live television
and radio interviews. He barely slept. "The course of least regret,"
he constantly repeated, "is to move to higher ground". And people
listened. More than 1 million US residents reportedly evacuated
their coastal homes.

Frank was not the only source of information about Gloria, of
course. Along with regular advisories from the NHC, regional
offices of the NWS issued local forecasts and warnings. Private forecasters employed by television and radio stations made their own
predictions. The American Telephone and Telegraph Co. teamed
up with NBC-TV to offer a special toll call telephone number for
periodic updates on Gloria's position.

Although the eye of Gloria passed over the Outer Banks of
North Carolina and across western Long Island, New York, to hit
the mainland's coasts, the hurricane's strongest winds remained
over water. By the time Gloria moved across eastern Canada and
back over the open waters of the North Atlantic as an extratropical storm – eventually producing record high temperatures over
parts of western Europe – many people were expressing disap-
pointment with both Gloria and the forecasters, charging them

with "crying wolf". The hurricane specialists defend their performance, however. Using their 24-hour forecast, the meteorologists said that the average error between Gloria's actual path and that predicted was 113 miles (181km). That compares to an average error of 127 miles (204km) for all 11 storms in 1985, and an average of 123 miles (197km) for all storms over the past ten years.

Why does the actual storm track vary from that predicted? Senior hurricane forecaster Gilbert Clark explained that the eye of a hurricane is often 20–30 miles (32–48km) wide and wobbles as it moves according to whimsical steering currents. Thus, a small error in locating the storm's precise center translates into a larger error over the span of a 24-hour forecast.

Despite the wealth of technology now available, meteorologists do not expect major improvements in their ability to forecast the path of hurricanes. The oceans are too large, the variables too many. In many ways, Gloria was typical – a violent hurricane fraught with potential for death and damage. Everyone knew it was coming. But where it would strike, and what it would do, was a mystery that no one could solve.

Hurricane Gloria bows out after threatening but eventually sparing the heavily populated East Coast. This satellite photo, taken September 29, shows the storm heading towards the open waters of the North Atlantic. Did the forecasters get it wrong?

HOW WEATHER IS FORMED

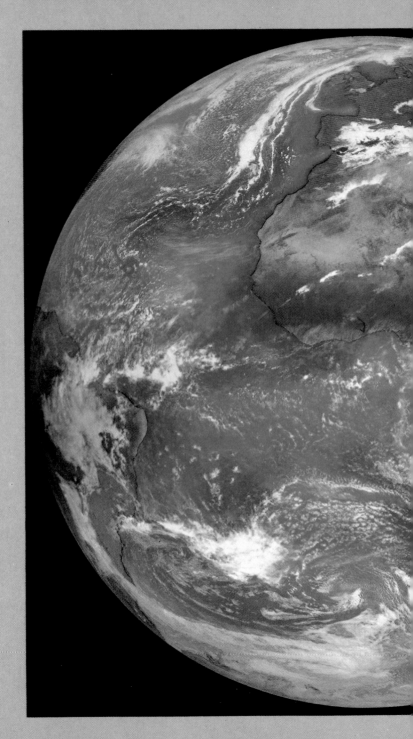

Cloud patterns reveal a range of weather systems in this full disk image taken from a geostationary satellite. Notice the trailing frontal clouds over the North Atlantic, the intense cyclone near Madagascar, and the deep tropical clouds over Brazil and Central Africa.

Weather systems

A satellite view of a tropical cyclone showing the bands of cumuliform clouds spiraling into the center. The storms result from very warm, moist air rising to great heights in the atmosphere over the tropical oceans. They are most frequent in the Atlantic and west Pacific oceans.

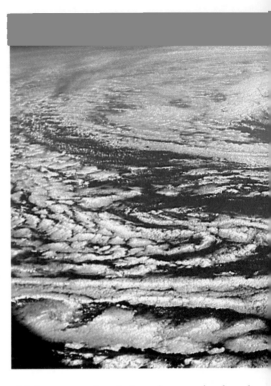

The human body's sensitivity to weather is largely perceived and expressed in terms of hot or cold, rain or sunshine, clear or cloudy skies. Thus the elements of the weather that we experience include temperature, sunshine, precipitation (all the forms of ice and water that fall from the sky), winds, humidity, mist, fog and clouds. These elements occur within the context of the weather systems, which are themselves mechanisms and processes of various sizes representing the atmosphere in motion. Weather systems form the essential framework for the weather elements, and these systems are ranked in a hierarchy of sizes, ranging from so-called Rossby waves that collectively cover half the planet, to *microscale* circulations the size of a football field. Among these, the *mesoscale*, or medium-sized, circulations are the most immediately responsible for changes in the weather elements. But these meso-scale circulations are themselves contained within another set of systems, called the *coriolis scale*, which includes cyclones and anti-cyclones. The latter in particular have long formed the bedrock of weather analysis and forecasting.

Maps on which weather observations are plotted are known as *synoptic* because the observations are taken simultaneously and "seen together" over a broad area – which is the literal meaning of synoptic. The snapshot facet of these weather maps provides their value, and the large areas covered by them gives a spatial meaning to the word synoptic. Hence it is now commonplace to use synop-

tic to describe the coriolis scale – the usual scale for weather maps presented on television. There have been many changes in operational weather forecasting during the last two decades, but an analysis of the weather elements can still usefully take place within the context of the coriolis scale with its familiar concepts of anticyclone and cyclone.

Cyclones

Undoubtedly the most familiar of weather systems are the cyclones and anticyclones. These are seen daily on the TV weather forecasts and in the newspapers as the most familiar effect of the weather experienced by the individual, since they usually express "good" weather (anticyclones) or "bad" weather (cyclones).

Cyclones are areas of low pressure around which winds blow counterclockwise in the northern hemisphere and clockwise in the southern hemisphere. Anticyclones are areas of high pressure around which winds blow clockwise in the northern hemisphere and counterclockwise in the southern hemisphere. The sense of rotation around these systems is a direct result of the coriolis effect (see p. 59) reflecting the Earth's spin.

Cyclones are of two main types: tropical and extratropical. Tropical cyclones (hurricanes) originate within, and largely travel within, the tropical belt. They are found most frequently over the western parts of the tropical Atlantic and Pacific oceans in the

1 Ice
2 Water and ice
3 Water

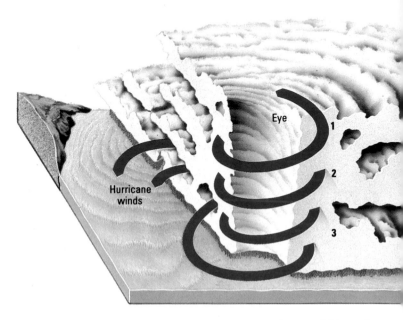

Eye

Hurricane
winds

northern hemisphere. In the Atlantic they originate within a few
degrees latitude of the equator, frequently in the eastern half of the
ocean. As they grow they move westward on the prevailing
easterly winds of the tropics. On reaching the Caribbean most of
them turn north and hence may visit the southern United States,
usually east of the Mississippi. Fortunately, many of them never
make landfall and travel northward over the western Atlantic to
die out around latitude 40°N.

The weather associated with tropical cyclones has been
described in the preceding chapter on Hurricane Gloria. The most
dramatic aspects are very high wind speeds and intense precipitation. The winds are closely related to the pressure distribution
(see p. 58). When a tropical cyclone passes over, the pressure may
fall by 50mb (millibars) in about 20 hours to a minimum of about
960mb, dramatic figures for both the size and speed of the change,
indicating sensational wind speeds.

Many satellite pictures of hurricanes show a familiar spiral of
cloud and look rather like pictures of galaxies and catherine
wheels. A typical storm is about 930 miles (1,500km) across and
the spirals are made of deep cumulus clouds. At the higher levels in
the storm, the rising air flows outward and eventually sinks on the
outer flanks of the system. It is the inflow and uplift of warm, moist
air that causes the high wind speeds and intense precipitation. In
the eye of the storm the opposite occurs. Winds are light and skies

A tropical cyclone seen in schematic vertical section. The bands and layers of cloud are shown spiraling around the center known as the eye. Within the bands the air rises rapidly. leading to deep clouds, heavy rainfall and high winds. Within the eye air sinks, resulting in calm, clear weather.

are clear. This results from the sinking air in the eye, which is only a few miles across.

Extratropical cyclones

Although far less savage than tropical cyclones, extratropical cyclones are larger, last longer, occur more often and affect the weather over much larger areas. They occur most frequently in the higher mid-latitudes, in the southern hemisphere between 50° and 60°S with little seasonal change, and in the northern hemisphere centered at 50°N in winter and 60°N in summer. In the southern hemisphere the belt of highest cyclone frequency is continuous around the globe. In the northern hemisphere the distribution of land and sea has some influence, resulting in the highest frequencies being in the North Pacific (the so-called Aleutian Low) and the North Atlantic (the so-called Icelandic Low). It is important to realize that the Aleutian and Icelandic lows are statistical rather than real. They are simply the areas through which cyclones pass most frequently – not areas where one cyclone persists as a permanent feature.

The extratropical cyclone is perhaps the aspect of meteorology most familiar to us. In one guise or another it appears every day on television and in the press. Typically, it tends to be a few thousand miles wide, as deep as the *troposphere* (the lowest level of the atmosphere, some 30,000ft high), with low pressure at its center and

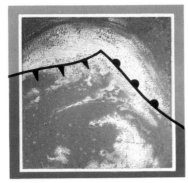

Fronts superimposed on satellite imagery illustrate the evolution of an extratropical cyclone. At **Stage 1,** the polar front has been buckled so as to give two components — warm (with blobs) and cold (with spikes). The cloud ahead of the warm front results from uplift at the front.

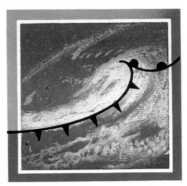

Between the two fronts lies the warm sector. **Stage 2** shows a well-developed cold front at the leading edge of a band of cloud that results from uplift at the front.

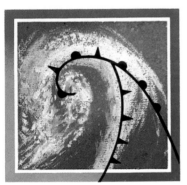

Stage 3 illustrates the occluding stage when the warm and cold fronts merge. The warm sector is lifted clear of the ground and accordingly diminishes. The cloud bands strikingly illustrate the cyclone structure at

this stage. The final stage **(Stage 4)** shows the demise of the cyclone when the depression is weakened and disperses, the fronts die, and the cloud pattern becomes disorganized.

therefore cyclonic circulation (hence its name), and almost always contains fronts.

Fronts were discovered around 1920 as lines of cloud and rain, and it soon became apparent that they were also zones of strong horizontal *temperature gradients* – boundary lines separating colder and warmer air. As such they provided the boundaries to the fairly uniform air masses described in the introduction.

At the same time as meteorologists began to understand fronts, they proposed a description of the life history of a typical cyclone which to a large degree remains true. The diagrams show how the front, which separates warm and cold air, is bent into a wave form as the cyclone matures. As a result, cold and warm portions of the

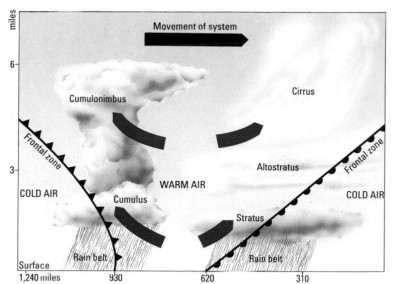

miles

6

3

Movement of system

Cumulonimbus

Cirrus

Frontal zone

COLD AIR

Altostratus

Frontal zone

COLD AIR

WARM AIR

Cumulus

Stratus

Surface

Rain belt

Rain belt

1,240 miles   930   620   310

A mature extratropical cyclone, seen in cross section. The cyclone is about 1,250 miles across and 7.5 miles deep. The cold and warm frontal zones separate different air masses and encourage uplift. The thin arrows show the airflow relative to the fronts and the stippled areas are clouds. Precipitation falls from the deep clouds around the surface location of the fronts.

front emerge, the former being the "boundary" when cold air replaces warm for the individual on the ground, the latter when warm replaces cold. The "blobs" and "spikes" we see on television screens are internationally agreed symbols. Between these two types of front within the cyclone, the wedge of warm air is known as the warm sector. The figure describes a vertical cross-section of the cyclone in its mature stage, showing how the cloud and precipitation are linked to the fronts. This scheme has had a very powerful influence on the analysis and forecasting of weather ever since it was first introduced more than 60 years ago. It is only in recent years that other techniques of forecasting have been used, as described on pp. 118–119.

The final stage of the cyclone occurs when the cold air behind the cold front reaches the cold air ahead of the warm front. This is known as an occlusion and in theory one of two things happens: the cold frontal air rises over the cold air ahead of the warm front; or it "undercuts" it. Despite the neatness of this idea, clear-cut examples in nature are rather elusive.

For many years the model of the frontal cyclone was accepted with virtually no modification. A major component of the model was the warm front with an associated suite of clouds and a large uniform area of precipitation (rain, snow, sleet, etc.) ahead of and parallel to the surface front. However, in the past 20 years improved techniques of observation have allowed meteorologists to find out more about mesoscale structure within both the fronts and other parts of the typical extratropical cyclone.

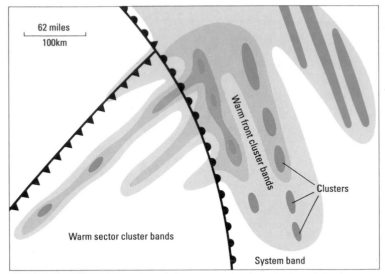

Rainbands within a mature extratropical cyclone (schematic map). Ahead of the warm front are three types of feature: the system band due to the overall uplift at the front; the cluster bands; and the clusters themselves. Cluster bands of similar size and shape occur in the warm sector, but here they lie parallel to the cold front.

Rainbands

Ahead of warm fronts, distinct bands of more intense precipitation occur (see diagram). These lie parallel to the front, about 30–125 miles (50–200km) ahead of it, and measure about 125 miles (200km) long and 30 miles (48km) wide. These rainbands are made up of clusters of "generator cells" about half a mile across, lying at heights of about three miles and carried along by winds at that level. This means that they move parallel to the front. The generator cells largely consist of ice crystals and result from convection, or vertical currents of air. This convection occurs as a result of cold, dry air from above the cold front overrunning warm, moist air carried over the warm front from the warm sector. This is an unstable state, so convection occurs. But why does it occur in lines or bands? With cold fronts this is because the front itself is a line, and the larger-scale airflow behind the convection forces a linear form. With warm fronts the situation is less clear, but the lines probably form in a "conveyor belt" of warm, moist air that flows parallel to the cold front before rising over the warm front.

All these mesoscale structures within the cyclones have their own distinctive sizes, durations and movements, so the accurate forecasting of the existence and timing of wet weather is particularly difficult. Until numerical forecasting techniques (see pp. 118–123) can cope with such small yet important systems, we must rely on close monitoring by radar and satellite, as described on pp. 109–110, which allows forecasters simply to estimate the paths of the rain areas.

This whole area of mesoscale structure within cyclones, particularly in relation to the distribution of rainfall and other types of precipitation, is currently one of the major research priorities of meteorology, since it is of great importance not only for members of the public but also for such services as hydrology, sewage and agriculture.

Anticyclones

Such is the concern with poor weather, which usually results from the passage of a cyclone, that anticyclones frequently get scant coverage. In fact, they occupy vast columns of the atmosphere and have profound effects, not always beneficial, on the weather of large areas.

In essence anticyclones are areas of relatively high surface pressure with sinking air and diverging, anticyclonic winds. Three types are recognized. First, subtropical highs, which are large, very deep systems, situated at about 30° on each hemisphere and roughly elliptical with the long axis along the latitudes. They are persistent features that stand out on a mean pressure map. Secondly, the polar continental highs, which occur over the northern continents in winter. They comprise very cold air, −13°F (−30°C) or colder, and are only about 1.5 to 2 miles (2.4 to 3.2km) deep. Thirdly, highs that form between cyclones, frequently as relatively thin wedges or *ridges*.

In all types the sinking air tends to inhibit, but not totally prevent, cloud growth, so anticyclones are frequently associated with clear skies and a good deal of sunshine. However, this does not always happen; low cloud and fog can readily form, sometimes giving an "anticyclonic gloom." The highs can also cause problems in winter when a *block* (see p. 40) may result in very cold air being drawn from high to lower latitudes or from very cold continental interiors to the usually milder oceanic fringes.

For many years, extratropical cyclones and anticyclones were considered to have separate origins. In fact, one theory neatly links them. The vertical motion of air is affected by processes occurring within the horizontal winds that sweep across the planet. If the horizontal area occupied by a given mass of air near the ground decreases (as the front end slows down and the back "catches up"), *convergence* is said to occur, causing an upward movement of air. If there is a horizontal spreading, or widening, of surface flow, *divergence* leads to downward movement. Such convergences and divergences, described on p. 61, occur in circulations covering huge areas of the planet, called Rossby waves; convergence is found on the upwind side of a *trough* and divergence on the downwind side.

Consequently, areas of surface high pressure – anticyclones – are found beneath the upwind high-level convergence, and areas of low pressure, or cyclones, are found beneath the downwind high-level divergence. These links also provide an excellent demonstration of the relationships between weather systems of differing scales (planetary to cyclone/anticyclone) and horizontal and vertical motion of the air.

Mesoscale circulations

Mesoscale weather systems are usually between 6 and 60 miles (10 and 100km) across. Two main types exist: those caused directly by interaction with the Earth's surface, and those whose origins owe little to the Earth's surface. In each type a further subdivision is made between systems due to *mechanical* effects and those due to *thermal* (i.e. heat) effects.

The mechanical effect of the Earth's surface is due to the presence of hills and valleys which are known as the *orography*. At the mesoscale, three main types of flow occur; *lee waves, downslope winds* and *circulations in wakes*.

Lee waves are just that; waves created on the lee or wind-free side of the obstacle that disturbs vertical flow. They tend to have wavelengths of about 6 miles (10km) and a crest-to-trough distance of a few hundred yards. They may occur in wave trains, usually a few tens of miles long, but occasionally a few hundreds of miles long, as has been observed downwind of the Andes. The waves tend to be best developed in the lower atmosphere, but they can be seen at all heights in the troposphere. The waves are stationary relative to the ground, but the air moves through them at perhaps 10mph (16kph). Frequently the air rising into the crest of a wave leads to *condensation* of water vapor and to striking forms known as "lenticular" clouds because of their lens-like shape. If several of these clouds exist one above the other, they give the impression of a pile of plates. Dramatic skies can result from lee wave occurrence.

The waves are due to air oscillating vertically as it moves horizontally. The horizontal motion is due to the general *pressure gradient* force, whereby air moves relative to the Earth (see p. 58).

Cloud streets (opposite) show as white bands over the eastern United States. Thermal convection causes typical cauliflower-shaped cumulus clouds which the wind then organizes into ''streets''. The spacing of the streets is determined largely by the typical breadth of a fair weather cumulus cloud.

Clouds form from condensation of water vapor in air that has been cooled by uplift; and airflow over mountains creates a major uplift mechanism. If the flow is stable, then as soon as the air crosses the mountain crest it will sink to its original level. If cloud forms in such a flow, it tends to cap the mountains, as shown here.

The lee sides of mountains often experience marked downslope winds, which may be warm or cold relative to the foot of the slope. These winds may result from the fact that when a stable airstream strikes a mountain and is forced to rise, it sinks immediately on the leeward side where the forcing ceases. If cloud and precipitation develop over a mountain, they would increase the warming effect of the subsequent sinking air. Such winds are characteristic of many of the world's mountainous regions, but the best known are probably the Föhn (Alps) and the Chinook (Rockies). The Chinook's ability to raise temperatures rapidly gives it the name of "snow eater".

Cool moist air

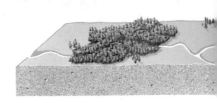

The vertical oscillation is due to the stability of the air. Forced away from its preferred height by the hill, the air tries to return to that height, overshoots, and again tries to return.

Downslope winds are either relatively warm (*Föhn*) or cold (Bora) and occasionally have devastating speeds of 90mph (144kph) or more. In both types, the downslope flow warms the air by *compression*, but the warmth or coldness of the wind is largely due to the temperature at the foot of the slope before the wind arrives. In central Europe and North America, particularly Alberta, the Föhns (Chinooks in Alberta) are warm relative to pre-existing air, whereas on the Adriatic, the customary mildness of weather conditions makes the Bora feel cold.

The origin of these winds, despite their being known for a century, is still not clear, but they are probably closely related to lee waves, the downslope wind being but the first part of a lee wave train.

Circulations in wakes result from the horizontal disturbance of flow around an obstacle in contrast to the vertical disturbance of lee waves. They form best in the wake of isolated islands such as Jan Mayen, but notable examples also occur downwind of the Aleutians and the Canaries. In shape they are alternately anticyclonic and cyclonic gyres, forming a widening train some hundreds of miles downwind of the island.

Thermally induced flows
Thermally induced mesoscale circulations near the Earth's surface are caused by heat (*thermos* is Greek for heat), and are due to the different cover of the surface – water, ice, bare earth, grass, trees,

Warm dry air

bricks and mortar and so on. Each type of surface has a different
response to the incoming energy from the sun and hence a different
effect on the temperature of the air above. The temperature affects
the air *density* and this affects the pressure, as explained on p. 46.
This basic mechanism applies to two important types of circulation; the sea/land breeze and the mountain/valley breeze.

Sea breezes blow from the sea to the land, being strongest in the
afternoon with speeds of about 11mph (18kph). They may occur
virtually every day over tropical coastlines, but they are sporadic
in middle latitudes in summer. *Land breezes* blow from land to sea
at night. They are far shallower and less intense than sea breezes.
Both types of breeze are part of a vertical overturning which
reverses from day to night. The temperature contrast between
water and land in the daytime results in relatively high pressure at
a height of about half a mile over the land. This drives air out to
sea at that height, leading to an accumulation of air which results
in relatively high pressure near the sea surface. It is this feature
which drives the sea air shoreward. The vertical circulation is
closed by rising air over the land, often at a quite sharp
"sea-breeze front," and sinking air over the sea. At night the
mechanism is reversed, resulting in the land breeze. In most cases
the sea/land breeze serves to ventilate otherwise stifling coastal
areas, but they are a mixed blessing if they merely recycle pollution emitted from coastal urban areas. Chicago and Los Angeles
suffer from this to some extent.

In hilly areas, air frequently moves up slopes in the daytime –
wisps of cloud can often be seen rising up a slope. This type of flow
is called *anabatic* (upward-moving). At night, air may drain down

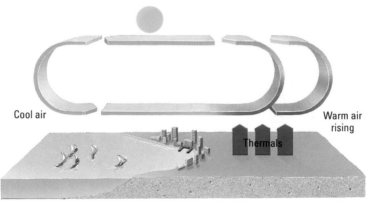

Sea breezes occur in daytime when air from the sea, being cooler and heavier than air overland, flows in to replace the rising warmer air of the coastline.

Land breezes occur at night and are due to the reversal of the land-sea temperature gradient. The land breeze is less intense than the sea breeze.

slopes – this is called *katabatic* (downward-moving) flow. In both, the cause is again basically heat. In daytime the tops of slopes or hills become warmer than air at the same height some way from the summits, because the summits absorb more solar *radiation*. This relative warmth is transmitted to the overlying air and this again leads to relatively high pressure over the summits. This drives air away from them and its accumulation over lower ground causes relatively high pressure near the ground, which drives air up the slope or valley. As with the sea/land breeze a vertical overturning occurs, which reverses its direction at night because of the relatively greater cooling at the hilltops.

The daytime flows help to diffuse any pollutants emitted in the valleys and also contribute to cloud growth over the hilltops – a frequent phenomenon, not least in the dry, mountainous areas of the United States. The nighttime flows lead to accumulation of cold air, which can result in frosts and fogs. In the special case of Antarctica, where the ice surface is perpetually cold, the katabatic winds can be severe enough to blow a person over.

Urban storms can be intense, giving inches of rain. It is likely that the extra heat, humidity and "roughness" of cities, compared to rural landscapes, can generate extra convective activity over the urban areas.

Storms and tornadoes

Some mesoscale circulations occupy the atmosphere above the boundary layer (the lowest layer of the atmosphere, about 3,000ft deep) and have origins virtually independent of the Earth's surface. The two main categories consist of severe local storms and circulations on cyclones.

Severe local storms occur over a range of sizes within the mesoscale, from tornadoes (a few tens of yards across) to mesoscale convective complexes (hundreds of miles across). It is their size, duration and internal structure that distinguish them from tropical cyclones and these characteristics, together with a different mechanism, distinguish them from extratropical frontal cyclones. Indeed, the severe storms form *inside* the frontal cyclones.

The basic building block of a severe local storm is a cumulonimbus cloud – sometimes known as a cell in this context. Some storms comprise but one cell which is about 3 miles (5km) across, 3–6 miles (5–10km) deep, lasts an hour or so, and has rising air at its core (the updraught) with speeds of 22mph (36kph) or more. The storm tends to move with the wind at heights of 2–3 miles (3–5km). It will give a shower of interim precipitation, possibly including hail.

Frequently, several such cells form a "multi-cellular" storm. Each cell evolves through its lifetime and the storm at any one time comprises young, mature and dying cells. Not only do the cells move, but their internal airflow is such that they tend to spawn new cells, usually on the front right of the main storm. Old cells die on the rear left. Hence the storm moves to the right of the wind that steers the individual cells. These storms are larger than the single cell, being 6–18 miles (10–30km) across and lasting for three to five hours or more.

Occasionally, one cell dominates a storm to become a "super-cell." It may be several tens of miles across, will certainly be as deep as the troposphere, will last several hours and will move to the right of and more slowly than the average wind in the troposphere. These storms can produce very large hailstones, such as the one weighing 1.7 lb (766g) that fell in Coffeyville, Kansas, on 3 September 1970. The diagram on p. 86 explains the mechanism of such thunderstorms.

Tornadoes originate in severe local storms, usually in the right rear quadrant. A tornado cyclone forms within the parent cloud and as the spin intensifies the tornado grows downward from the cloud. The highest windspeeds on Earth occur in tornadoes – so high that they devastate everything in their path. Tornadoes occur most frequently in the middle states of the United States, particularly from Texas north to Kansas. They are also found east of the Andes, in eastern India and, in very muted form, in the British Isles.

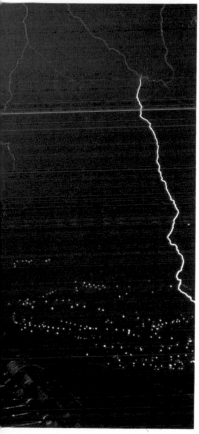

Hailstorms (top) and thunderstorms develop when cumulus clouds grow to depths of 6 miles (10km) or so. The clouds glaciate in their upper levels and become cumulonimbus. Within these large clouds the updraughts are such that hailstones grow by accretion of supercooled water droplets onto an ice ''embryo''. Hail falls in localized shafts which are clearly visible in the picture above. Frequently associated with hail growth is the separation of electric charge within the cloud. Break-up of ice crystals leads to positively charged fragments being carried to the top of the cloud whereas larger, negatively charged portions fall to the base of the cloud.

Lightning (left) results from these charge separations. The first stage of a lightning flash brings down negative charge from the cloud base and is met near the ground by a return stroke which takes positive charge upward along the already formed channel. Additional leads and returns may occur, but the duration of the total flash is about one fifth of a second. The expansion of the air due to the lightning sets up sound waves which we hear as thunder.

Microscale and local weather patterns

At the small end of the hierarchy of systems lie the *microscale* flows, where the systems are of the order of inches, yards, and, at the most, a few miles across, rather than the tens and hundreds of miles that we have been considering. Small air flows such as these are most often found in the lowest layer of the atmosphere, the so-called boundary layer, and as a result are strongly affected by the thermal (heat) and mechanical effects of the Earth's surface.

Urban areas display some of the most pronounced of these effects. Buildings provide a rougher lower boundary in terms of relief and contour than do most natural landscapes, and this has the effect of both channeling winds and therefore often increasing their speed at a very local level, and of reducing average wind speeds because of increased frictional drag. Urban areas may cause warm air to rise and, if it is sufficiently moist, to form cumulonimbus clouds. So thunderstorms and summer showers may be a feature of the urban weather. Urban and industrial areas are also major sources of pollutants. When an air mass conducive to fog formation and containing no rising air currents settles over a city, the result may be *smog*, which, as in Los Angeles, may dominate the perception of a city's weather. London, at one time subject to killer smogs, has greatly benefited from smoke controls introduced 30 years ago, leading to dramatic increases in visibility and sunshine hours.

Because urban areas act rather like storage heaters, towns often tend to be significantly warmer than the surrounding country at night and in the winter. This is known as the "heat island" effect. The materials from which roads and buildings have been constructed absorb heat during the day, releasing it at night. Moreover, drainage systems reduce evaporation, buildings block winds, and of course, a considerable quantity of heat is generated by factories, homes and offices. Thus large metropolitan areas such as New York can be 10°F warmer than surrounding rural areas.

Areas of ice also produce significant microscale circulations. Ice is far more uniform than an urban surface, although it takes different forms such as valley glaciers and large sheets. But whatever the type, the effects of the ice are the same. Because it has high reflective capacity or *albedo* and very low temperature, the usual decrease of temperature with height is inverted, causing strong downslope flows of air.

Wooded areas resemble urban areas in some respects, but they differ importantly in that wooded areas are alive and use water, which modifies the local water balance of soil and atmosphere – sometimes to the disadvantage of other users of land in the vicinity. Trees are often used as a shelter because they exert a frictional drag on air and thus reduce wind speeds.

All these effects we have just described are perhaps not to be thought of as weather in themselves. It is more the case that the Earth's surface tends to exaggerate a particular element (wind, for instance) within a given cyclonic or mesoscale circulation, or to change several such elements in a small way that, when added together, may result in a noticeable difference.

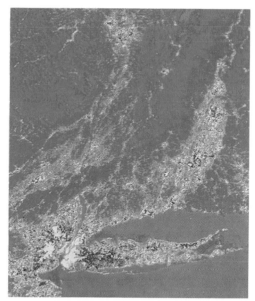

The heat island of New York from nearly 400 miles up is pinpointed by a satellite measuring radiation and hence, indirectly, temperature. The colors represent different temperatures, from red and white (warmest) to blue and violet (coldest).

Smog accumulates in stable air over sources of water and smoke. A prime area is the Los Angeles basin (below) where the terrain, atmosphere and millions of automobiles combine to produce an obnoxious brew, trapped by the surrounding mountains.

The general circulation of the atmosphere provides the stage for all weather and climate. It has been investigated for over 300 years, with the aid of only surface observations for much of the period. In the past 20 years it has been possible to see large areas of the globe from satellites. These images have revealed a bewildering amount of detail in the atmosphere, largely through the cloud patterns. As the picture shows, it is not always easy to identify the apparently simple large-scale airflow regimes from the typical cloud imagery.

Planetary scale systems

At the other end of the scale of weather systems, two planetary scale systems dominate the atmosphere; the *Rossby waves* that exist poleward of about 30° and the *Hadley cell*, a vertical overturning of the atmosphere in the region between the 30° parallels.

The Rossby waves are so called after Carl-Gustaf Rossby, the eminent Swedish-American meteorologist who provided the first theory of them. The waves are three-dimensional, up to six can exist, and at minimum they are several thousands of miles long, as are the crest-to-trough distances, extending over tens of degrees of latitude. The waves probably originate due to massive mountain ranges such as the Rockies and the Andes, but the distribution of land and sea may reinforce the pattern.

A major problem in meteorology is to understand the changes in both the number and shape of these waves. They take on myriad shapes, but two extremes are worthy of note. When the poleward temperature gradient over a hemisphere is strong, the waves tend to have small amplitudes, large wavelengths and to contain relatively high wind speeds. The latitudinal temperature and pressure gradients across the waves are strong, which encourages the rapid easterly movement of active depressions with important effects on the weather of areas affected by the waves. When the latitudinal gradients are weak, there is an increase in wave number and amplitude, but a decrease in wavelength. In turn these lead to "blocking," where high pressure systems act rather like boulders in a stream. This causes the extremes of weather, particularly in northwest Europe. In summer, high temperatures and dry weather can occur for several weeks, such as in 1976; in winter, very low temperatures, ice and snow may persist for months as in 1962-63.

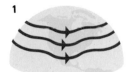

1

Rossby waves are three-dimensional and extend across latitudes. Stage 1 shows a strong latitudinal form where the

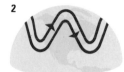

2

wavelengths are long but their amplitudes are small. This may change to one of the shorter wavelength and larger amplitude (stage 2),

3

eventually reaching a more extreme form (stage 3) that yields so-called ''blocking highs'' of fixed weather.

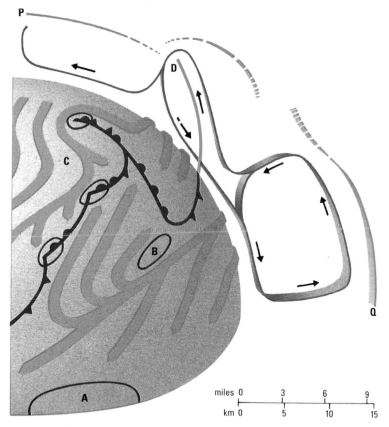

| miles | 0 | 3 | 6 | 9 |
| km | 0 | 5 | 10 | 15 |

Warm front

Cold front

A Area of low pressure
B Area of high pressure
C Area of low pressure

D Polar front
P Polar cell tropopause
Q Tropical tropopause

The general circulation of weather in the northern hemisphere, schematically shown. In the tropics, the Hadley cell is a vertical overturning that reaches much higher into the atmosphere than does that of the polar cell. In between is the Rossby regime of waves, within which extratropical cyclones form and within which, in turn, the fronts form. Bold arrows on the globe show surface airflow.

The Hadley cell

The Hadley cell is so called because the English scientist George Hadley first suggested that a cellular overturning of the atmosphere along the *meridians* could achieve the poleward heat transfer required to keep the heat of the atmosphere in balance (see pp. 48–49). In fact Hadley proposed that such a cell existed between 0° and 90° in each hemisphere, with warm air rising over the equator moving poleward at high levels, sinking over the poles and returning equatorward at the surface. Later observations showed that this does not happen, but that a cell operating in the sense Hadley described does exist between latitude 0° and 30° in each hemisphere. It is the rotation and curvature of the Earth that limits the latitudinal extent of the regime. Outside the tropics, a vertical overturning becomes unstable and collapses into the Rossby wave system already described.

The Hadley cell is the simplest kind of vertical overturning: warm air rises, cold air sinks and two "horizontal" arms complete the circulation. Near the equator the warm, moist, unstable air, largely over ocean of course, rises in a multitude of cumulus towers. The air contains much *latent heat*, which is released on condensation to warm the upper tropical atmosphere. At heights of 39,000–49,000ft (12–15km) the air is cooled by radiation, a process that continues as the pressure gradient drives the air poleward. At about latitudes 30° the air sinks in the subtropic high

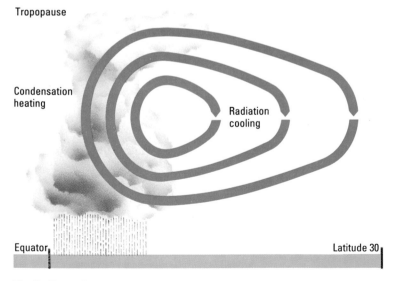

Tropopause

Condensation heating

Radiation cooling

Equator   Latitude 30

The Hadley cell equatorial convection system causes air to rise in the Inter tropical Convergence Zone, shedding rain. At the tropopause, the air moves poleward cooling by radiation, and eventually sinks over the subtropics. Low-level flow toward the equator then forms the trade winds.

Tropical weather can be seen in a satellite image, where cloud associated with the Intertropical Convergence Zone stands out clearly along the equator belt from West Africa and into the Atlantic. Lack of cloud over the Sahara desert is strikingly evident.

pressure areas, warming by compression (rather like a vigorously used bicycle pump) and inhibiting cloud growth. Hence skies are frequently clear, and the receipt of solar radiation and near-surface temperatures are both high. Here lie the great, hot, dry deserts of the world such as the Sahara and the Australian.

The Intertropical Convergence Zone

Because of the high pressures in the subtropical latitudes, and the low pressures near the equator, the Hadley cell is closed by massive streams of air flowing equatorward. The effect of the Earth's rotation, although weak at these latitudes, gives an easterly component to the winds which are familiar as the north-east (northern hemisphere) and south-east (southern hemisphere) trade winds. The area where the trades meet is called the Intertropical Convergence Zone. The average position of the zone shifts north and south with the seasons, in response to the overhead sun. Its annual average position is a few degrees latitude north of the equator.

The zone is not a continuous girdle around the Earth. It comprises large, linear segments, several thousand miles long, that frequently show clearly on satellite photographs. Within these segments are smaller systems, called in decreasing order of size, cloud clusters, mesoscale convective cells and individual convective elements.

Monsoons

The word *monsoon* comes from the Arabic *mausim*, meaning a season, and herein lies its nature and origin. Monsoons are circulations that, on average, reverse their direction every six months. The term was first applied to winds over the Arabian Sea and the major circulation of which these winds form a part, the Asian monsoon, is the prime example. But monsoonal circulations have been described in Northern Australia, parts of Africa, South America and the United States. In this section we are considering the situation in Asia.

The Asian monsoonal circulation is particularly well developed over the Indian subcontinent. In summer the near-surface flow sweeps over southern Asia from well inside the southern hemisphere, this being by far the most prominent interhemispheric atmospheric transfer on the planet. Near India the southwest monsoon brings heavy rainfalls and then sweeps around to become southeasterly up the Ganges valley. Above these low-level winds, the high-level tropical easterlies flow from the western Pacific to Africa, straddling the equator up to 15° or so of latitude. In winter the low-level winds dramatically reverse, being northeasterly over India and immediate surroundings, and northwesterly over Chinese Asia. But these low-level winds do not cross the Himalayas. The upper-level flow also differs from summer with the westerlies now overlying all but the extreme southern part of the monsoon regime.

The basic cause of wind reversals such as these is thermal. In many respects the monsoon is rather like a gigantic sea breeze. In summer, land heats up more than nearby water and, in the same fashion as is found in the thermal wind mechanism, this leads to a high pressure at high levels over land and low pressure at the same height over sea. Air moves from land to sea, creating a pressure gradient in the opposite direction at the surface. This drives air from sea to land. In winter the process is reversed because the land cools down much more than the sea in this season.

In the Asian monsoon this simple, thermally direct circulation does not provide a full explanation for the weather experienced. Both *jet streams* and the mountains and valleys of the area have important effects. In summer, the thermal mechanism indeed operates, but the progression of the southwest monsoon and its passage up the Ganges valley is hindered by the existence of a westerly jet stream lying south of the Himalayas. The sheer height of these mountains inhibits the usual poleward shift of this jet, but when it happens, over a period of a few days, the monsoonal air "bursts" in over the Indian subcontinent. Some of the highest precipitation amounts in the world are recorded in the Assam hills. In winter the Tibetan massif cools, creating a shallow, cold layer of surface air and shifting the continental cyclone aloft toward lower latitudes. The Himalayas once again have an important effect, preventing the simple "overflow" or outflow of cold air from the continental interior. Subsidence under the westerly high-level jet stream appears to play a vital role in the production of the surface northeasterly winds that give a warm, dry winter.

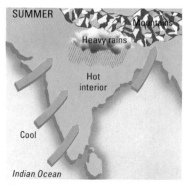

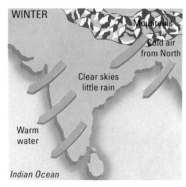

The monsoon in summer over the Indian subcontinent begins as monsoonal air approaches from the Indian Ocean. This airflow brings some of the world's heaviest rains, particularly in the Assam hills.

The winter or northeast monsoon occurs when the wind reverses, with cold dry air from the north flowing southwestward from the Asian interior. This period tends to be relatively dry.

The monsoon is not restricted to the Indian subcontinent. Monsoonal circulations occur over much of southeast Asia, including China and over parts of Africa. The picture shows monsoonal rains in Indonesia.

The Earth and the Sun

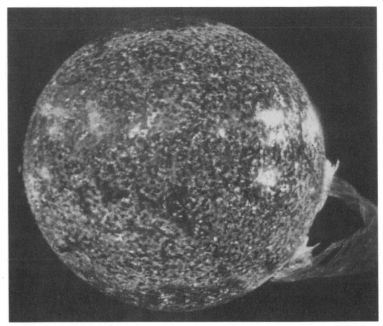

The Sun is the primary source of energy for the Earth. Our planet nonetheless intercepts only a small fraction of its total emission.

The fundamental cause of all weather on Earth is the Sun and its position relative to our planet. This is not just in terms of the various seasonal changes experienced as the Earth progresses through its yearly orbit. Heat energy supplied by the Sun directly affects the density of air (hot air is lighter than cold air) thus setting up the all-important pressure gradients which cause air to move in an attempt to minimize gradients in its distribution. The constant motion of the atmosphere is thus dependent on energy balance, and this needs to be considered from two points of view: the balance – or budget – between Earth and space, since this determines the average temperature of the atmosphere; and the balance or budget within the atmosphere itself, since this is the basic cause of all weather.

The Earth-space "budget": input

All budgets are a matter of income and outlay, or input and output. In this case the input is the radiation received from the Sun; the output is the radiation from the Earth. In the long term these quantities should balance, but in the course of the Earth's history it is thought that slight imbalances may have taken place, as evidenced perhaps by the occurrence of ice ages.

The Sun sends out short wave radiation at a rate that varies little – hence the name Solar Constant. This provides the energy for all natural life and movement on our planet. When it reaches the Earth, this solar radiation is reflected, back-scattered and

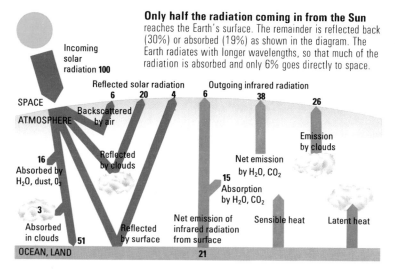

Only half the radiation coming in from the Sun reaches the Earth's surface. The remainder is reflected back (30%) or absorbed (19%) as shown in the diagram. The Earth radiates with longer wavelengths, so that much of the radiation is absorbed and only 6% goes directly to space.

absorbed by various components: 6% is scattered back to space by air itself; 20% is reflected by clouds, and a further 4% by the Earth's surface. Hence 30% of the radiation is lost to the planet by these processes, which collectively comprise the albedo. Clouds absorb 3% of the remaining solar radiation, while water vapor, dust and other components in the air account for a further 16%. The result of all these atmospheric interferences is to ensure that only 51% of the incoming solar radiation actually reaches the Earth's surface.

This figure is only an average, and it conceals the variations in the quantity of solar radiation that reaches the ground at different points on the planet. Since the Earth is spherical, three times as much solar radiation reaches the tropical areas as reaches the polar regions. Moreover, due to the distribution of cloud cover, the equatorial areas receive only half as much solar radiation as reaches the hot dry deserts of the world, where some 80% of the total radiation penetrating the atmosphere reaches the ground. And in the cloudy middle latitudes, the solar radiation receipt at the ground is only one third of that found in the deserts.

The Earth-space "budget": output
The solar radiation input has to be balanced by an equal output of heat back from the Earth. This is achieved by radiation from the atmosphere.

Unlike the short wave radiation, Earth radiation occurs in long wave form and is thus far more readily absorbed by the water vapor and carbon dioxide in the atmosphere. Of the radiation emitted by the solid Earth, some 90% is absorbed by the atmosphere, which radiates about 80% back to the Earth. Thus the atmosphere acts like a blanket or like the glass of a greenhouse – hence the so-called *greenhouse effect*. As a result, only a very small amount of radiation from Earth escapes directly into space.

Energy transfer within the atmosphere

The atmosphere receives heat from short wave solar radiation, long wave Earth radiation, and additionally from convection, whereby vertical currents of uprising air liberate heat energy (both *sensible* and *latent*) from the Earth's surface. The atmosphere loses heat by radiating upward to space and downward to Earth. In fact, the atmosphere would have a negative heat balance, losing far more than it gains, where it not for the fact that the deficit in radiation is made up by the influx of heat by convection. Without this convection, the Earth's surface would have to be much hotter – about 152°F (67°C) instead of 59°F (15°C) in order to emit enough radiation to balance the budget of thermal (heat) equilibrium.

Satellites can now tell us how this long wave radiation is distributed using measurements from *infrared* imagery. Observations show that amounts are highest in the desert areas, lower in the middle latitudes, and lowest in the polar regions. The resultant gradient of output, though relatively small, has important repercussions on the need for weather systems.

Balancing the energy budget

We have seen how the Earth maintains an energy balance with space, with the atmosphere playing a critical role in this budget. We must now consider how the energy (or heat) budget works within the atmosphere system since, as we have seen, this is the basic cause of all weather.

Leaving aside the role that convection plays in this transfer of heat, we should consider the balance between the incoming and outgoing radiation, known as *net radiation*. Here it is important to look at the distribution of net radiation both for the Earth's surface and for the atmosphere. These considerations lead in turn to the net radiation for the entire Earth-atmosphere system, which is one of the most important distributions in the whole science of meteorology, and indeed of natural science on Earth.

Over all latitudes between 80°N and 80°S, the Earth's surface has positive net radiation – that is, it receives more than it loses. This positive net radiation is particularly high in the tropics. The atmosphere, on the other hand, has a negative net radiation balance, which is constant with latitude. Adding together the two distributions makes up the net radiation of the Earth-atmosphere system.

Between the latitudes of about 40°N and 35°S the net radiation budget is positive. Poleward of these latitudes the budget becomes negative. Unless there were some other factor affecting the radiation budgets, this distribution suggests that the tropics would get progressively hotter and the poles progressively colder: the equator would be 25°F (14°C) warmer and the north pole 45°F (25°C) colder than they in fact are. Since this is not the case, there must be a transfer of heat from the tropics to the poles. Such a transfer, which is known as the *meridional flux*, ensures that the atmosphere's temperatures are remarkably stable, and that the gradient of average temperature across the latitudes is some 72°F (40°C) less than it would otherwise be. In addition to this meridional flux, heat is also transported upward: the net effect being that the variations across latitudes of long wave radiation at the "top" of the atmosphere are much smaller than those of the incoming solar radiation.

The necessary transfer of heat toward the poles (the meridional flux), takes place in both atmosphere and ocean, with the former accounting for about two-thirds of the total. The transfer is achieved through the medium of moving air: warm, moist air moves upward and toward the poles; cold, dry air moves downward and toward the equator. The configurations of air flow that bring about these movements are essentially the weather systems of our planet Earth. The need to balance the net radiation budget is the reason why weather exists at all. This is the critical link between the long-term climatic characteristics of the atmosphere and the turbulent motions, active at many different levels, that provide our weather from day to day. The distribution of net radiation within the atmosphere is the driving force behind the Earth's major weather systems which have already been described in an earlier chapter.

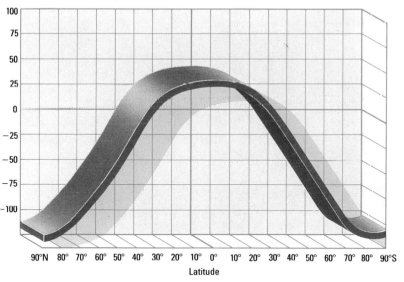

Latitude

Net radiation is the difference between the incoming and outgoing radiation. At the Earth's surface much more radiation is received than lost, so the net radiation is positive over virtually all latitudes. But the net radiation balance of the atmosphere alone is negative over all latitudes. Adding these two distributions gives the curve shown here. For the Earth and atmosphere together the net radiation is positive between latitudes 40°N and S and negative elsewhere. It is this distribution which drives the atmosphere and oceans, constantly seeking to reduce the latitudinal gradient of energy resulting from the net radiation.

Sunshine and temperature

Sunshine, the light directly received from the Sun, occurs at the ground in the absence of cloud. Meteorologists measure it in terms of numbers of hours rather than intensity. The highest amounts of sunshine are recorded in the areas beneath the zones of subtropical high pressure that make up the world's hot deserts. In the extratropics (the areas of the Earth outside the tropics) abundant cloud cover may mean that many days are completely without sunshine at the surface. This seems to be depressingly true of the northwestern coast of the northern continents, where many winter days are dominated by cloud.

Sunshine clearly affects temperature on Earth. We have seen, however, that the atmosphere as a whole is primarily heated from *below* – by long wave radiation from the Earth, and by convection. *Conduction* – the transfer of heat through direct contact – is also involved to a lesser extent.

Convection is a daytime feature, occurring when heat is moved from surface to atmosphere by local volumes of warmer and less dense air known as thermals. At night, convection is absent and conduction and radiation take over as major vertical transporters of heat. Conduction is effective only in very still air and then only

Fresh snow may reflect back over 90% of sunlight received, which clearly means that very little heat gets into the surface. It is this high reflectivity which can cause snow blindness and gives the very low temperatures found on clean snow surfaces. It is warm air, not sunshine, that primarily causes snow to melt.

Major hot desert areas (right) give the highest values of solar radiation as measured at the ground. There is little water to take heat for evaporation and little vegetation to use water and shade the ground. Hence the radiation is absorbed by the surface, giving some of the highest air temperatures found on Earth. Surface temperatures are even higher.

over a few inches. Long wave radiation loss by the surface leads to cooling, and this is transmitted to the overlying air. This leads to cooler air near the ground than at higher levels. The increase of temperature with height is called an *inversion* – it is the opposite of what usually happens – and in this case it is caused by radiation.

Some very low temperatures can result from this mechanism. The lowest surface temperatures on Earth are at the Pole of Inaccessibility in Antarctica where they fall to −112°F (−80°C) or less. Temperatures of −40°F (−40°C) frequently occur in North America and Siberia in winter, when anticyclones bring cold, dry air from the polar zones. Such a dry atmosphere allows the long wave radiation to escape easily, with practically no warming effect. A further factor contributing to coldness is fresh snow cover. Fresh snow has a thermal conductivity only about 5% that of soil, so when its surface loses heat by radiation there is little replenishment from below. Hence its surface temperature rapidly falls. A final factor may be the downward movement of the air, which leads to the accumulation of a "lake" of very cold air. So you can expect low temperatures in conditions when the skies are clear, the air at the surface is calm, fresh snow has fallen, and there is a dry atmosphere and downward-moving air.

High temperatures are also associated with high pressures for several of the same reasons listed above for low temperatures. The main differences are the absence of snow and the presence of solar radiation. Otherwise, clear skies, calm air and a dry atmosphere all favour high temperatures. In such a situation solar radiation is unimpeded in its path to the ground, where it is absorbed, giving very high surface temperatures (egg-frying heat). These in turn lead to high near-surface air temperatures because the calm air remains in the same place to receive the heat. In August 1981 the far west of the USA had a record-breaking heatwave with temperatures about 9°F (5°C) above average in the first week or so of the month. This was basically due to the massive subtropical high extending well into western USA allowing very warm dry air to build up. Record temperatures included 108°F (42.2°C) at Eugene, Oregon and 121°F (49.4°C) at Red Bluff, California.

In the above examples of extreme heat and cold the wind speeds were very low, allowing radiational processes full play. When winds occur, two major things happen. First, air does not stay in one place long enough to take on any extreme characteristics the underlying surface may have. Secondly, whatever character the air does have is carried by the wind, and although modified somewhat on its track, strongly influences the temperature of its destination. Indeed this *advection* of air is a major determinant of the temperature at a given place. For example, northern bursts of air east of the Rockies lead to cold spells in the southern states, sometimes damaging the orange crops in Florida. Conversely, northwest Europe's winters are generally so mild because southwesterly winds bring relatively warm moist air from lower latitudes. Once more the paramount importance of air motion is abundantly clear.

Water may appear to "smoke" or steam in high latitude oceans, as shown in this picture taken on northwest Greenland. This results from the large temperature contrast between the water and the air. The water is relatively warm and the air is very cold and dry. Hence water evaporates from the oceans into the dry air, only to be condensed almost immediately due to the coldness of the air. The net result is a dramatic "steaming" effect.

Drought occurs when, over a period of weeks to months, more water leaves the surface of the Earth than reaches it through precipitation. The result may well be as shown below, a parched, deeply cracked land that is useless for sustenance. In recent years the Sahel and Ethiopia have experienced such conditions, leading to massive sympathetic responses such as the 24-hour rock extravaganza Live Aid.

The nature of the atmosphere

The atmosphere, the home of our weather, may now almost all be seen by means of satellites. The apparent chaos of the atmosphere, shown here in the multitude of cloud forms, hides an order which meteorologists continue to seek.

The weather that we experience is formed in the atmosphere – the ocean of air that surrounds our planet. What is the actual nature of air, and how is it distributed around the Earth?

The Earth probably had no atmosphere when it was formed over four and a half billion years ago. As it cooled, gases such as carbon dioxide, nitrogen and water vapor were released from the surface. Then heavy rains, falling for long periods, washed out most of the carbon dioxide, simultaneously creating vast oceans and lakes. Oxygen, without which no air-breathing animal could survive, came later, due perhaps to the breakup of water molecules by radiation or the activity of primitive algae. This life-giving gas was further increased by plant photosynthesis (the process of absorbing sunlight and carbon dioxide from the air, and water from the soil, so as to give out oxygen).

So air is not a single gas but a mixture of several gases. Air that contains no water vapor is made up of nitrogen (78%), oxygen (21%) and argon (0.9%), with other gases such as neon, helium, methane, krypton and hydrogen making up less than 0.002%. There is also carbon dioxide, which has a concentration of only about 0.0335% but has an important effect on the temperature of the atmosphere. For this gas lets the Sun's radiation through to the Earth, but absorbs Earth's outgoing radiation. The massive consumption of the planet's fossil fuels over the past 30 years has led to a significant increase in carbon dioxide, which could produce a small but potentially catastrophic warming of the atmosphere, with a massive rise in sea levels. Fortunately other factors lead to a compensatory cooling. But non-gaseous constituents, such as very small dust and smoke particles, are also important as absorbers and reflectors of sunlight, and hence influence temperatures.

Carbon dioxide is a variable gas – its quantity fluctuates according to conditions. Another variable gas of even greater significance is water vapor, which can range from 0% to 4% by volume.

The weight of air

The various gases of which the air is composed, and which make up the atmosphere, have weight. Thus air exerts pressure at the Earth's surface which, at sea level, is about 14.7 lb per sq.in. In the metric system internationally used by meteorologists, this atmospheric pressure is expressed in millibars; 1 millibar equals 0.03 lb per sq.in., so the average sea level air pressure is 1013.25 millibars.

Because air can be compressed, the weight of the atmosphere is not evenly distributed around the Earth. Its vertical distribution is also uneven. About half the atmosphere lies below a height of 3 miles (5km) and almost 95% lies below a height of 18 miles (30km). Pressure, density and temperature all change with height, forming four distinct temperature layers, each separated by a surface or "pause." It is the behavior of the atmosphere in the lowest of these layers, the troposphere, that largely determines the weather that we experience from day to day.

The average state of the atmosphere

Atmospheric pressure, temperature, winds and humidity make up the background of the weather. But, as we have seen, the average pressure significantly decreases with height; and there are also significant horizontal variations, particularly at the Earth's surface. Both the northern and southern hemispheres have a pattern of low pressures near the equator, high pressures in the areas near latitude 30° (the subtropics), low pressure at latitude 60° or more, and high pressure over the poles. This broad latitudinal pattern is remarkably persistent, apart from a major seasonal change that occurs over Asia where high pressure exists in winter and low pressure in summer.

Surface temperatures also have a global pattern. A broad gradient runs from high temperatures at the tropics (77° to 86°F or 25° to 30°C) to low ones near the poles (−4° to −40°F or −20° to −40°C), with temperatures decreasing with increasing latitude. Smaller-scale variations occur, however: in January the North Atlantic is abnormally warm for its latitude, whereas Siberia and Canada are abnormally cold. These apparent irregularities are due to the warmer waters of the North Atlantic and the colder land surfaces that border it. In July these anomalies disappear and the extremes are found in the Sahara and Antarctica.

Effects related to the Earth's surface can lead to the formation of "air masses," or large parts of the atmosphere where temperature and humidity vary very little at the horizontal level. The cold landmass of Siberia, for instance, produces continental polar or arctic air that is also very cold and relatively dry. The warm waters of the oceans, on the other hand, produce maritime tropical air that is mild and moist, a major characteristic of the weather

Lowest near-surface temperatures are found over snow surfaces at high latitudes, as here in Greenland. In winter temperatures fall to −22°F (−30°C) or lower. Near-surface temperatures are highest in the subtropical hot deserts.

and climate in the north west of Europe and of the United States.

The distribution of surface winds follows that of surface pressure, with two main patterns in each hemisphere. In the tropics the trade winds dominate, blowing from northeast in the northern hemisphere and southeast in the southern hemisphere, in response to the pressure gradient between the subtropic highs and the equatorial low.

In the middle and high latitudes the average winds are essentially westerly around the globe in both hemispheres. The two basic westerly and easterly patterns exist throughout the depth of the atmosphere over appropriate latitudes.

Average humidities fall with both latitude and height. The air is 60 times moister near the surface in the tropics than it is three miles above the polar regions.

The boundary layer

The *boundary layer* of the atmosphere is generally agreed to be the bottom 3,000ft (1km). It is here that much of the energy and all of the water and momentum enter the atmosphere: hence its importance to the general circulation. It is also, of course, the part of the atmosphere where much of the pollution enters and accumulates, where turbulent eddies are at their most complicated, and where humans live. Unfortunately, it is also one of the most difficult parts of the atmosphere to understand.

Heat and water enter the atmosphere at its bottom or boundary layer largely by natural processes. But industrial activity may add to this transfer, as seen here with a cloud developing from the cooling towers' hot, moist effluent.

The boundary layer is divided into three distinct vertical regions. The surface layer, a few hundreds of feet deep, has strong gradients of wind, temperature and humidity. Above this is a mixed layer which occupies most of the boundary layer depth. Because of the mixing, the average vertical gradients of wind, temperature and humidity are small. Finally, a layer where temperature increases with height caps the boundary layer as a whole. In daytime the boundary layer is frequently full of turbulent eddies which are driven by convection. If they are intense, we will experience sharp gusts of wind. If they are less marked, the eddy structure and mixing is more likely to be evident only in the behavior of smoke, such as emerges from industrial chimneys.

At night the boundary layer air cools. This damps down turbulence and may allow waves to form in the layer. Mixing is drastically reduced and this further encourages the development of strong vertical gradients. If skies are clear, temperature at and near the surface may drop sharply, possibly leading to frost or fog formation. Valleys encourage the drainage of the cooled air and tend to have the most severe frosts and thickest fogs.

Pollutant dispersal is dependent upon the characteristics of the boundary layer. In siting potential emitters it is important to assess several factors, including the local structure of the layer, the type, amount, frequency, and height of release of the emissions, and the local terrain.

The air in motion

Winds are simply bodies of air moving relative to the Earth. The Earth itself is rotating, of course, and at the middle latitudes the surface is moving from west to east at about 738mph (1,188kph). So a westerly, or wind blowing from the west, at 22mph (36kph) as measured from the Earth, is in fact moving at 760mph (1,224kph) as measured from space. Easterly winds are bodies of air moving from the west at speeds slower than the surface speed. The word "wind" is generally used to describe horizontal flow; vertical motion also qualifies as wind, though the term "current" is widely used. In fact, as we all know from experience, winds tend to gust and lull, largely in response to their complicated structure of eddies, or whirls of air somewhat similar to the whirlpool of water at the open outflow of a bath. Instruments measure the effects of these eddies on a timescale of seconds, but for practical purposes the record is smoothed over minutes for both speeds and direction. So to understand the nature of winds it may be helpful to separate them into horizontal and vertical flows.

Horizontal motion
Air resembles water in that it is a fluid, or flowing, medium, and in many respects it behaves in a similar fashion. Thus it moves horizontally relative to the ground only if a force is applied to it. This force results from the horizontal differences in atmospheric pressure, which themselves result from the uneven distribution of the atmosphere. The force is therefore called the *pressure gradient force*, and the stronger the gradient – the "slope" between high and low pressure – the stronger the wind.

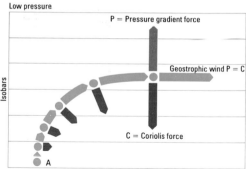

Low pressure

P = Pressure gradient force

Geostrophic wind P = C

Isobars

C = Coriolis force

A

High pressure

Cloud-forms in a ragged sky (left) trace the pattern of a turbulent airflow. Wind is air moving relative to the Earth. But as air is a fluid, the flow is far from simple and may be highly complex. The water content of the atmosphere means that we can frequently see this turbulent flow in the cloud-forms.

The geostrophic wind (above) results from a balance of the pressure gradient and coriolis forces. Here the air particle at A moves from high to low pressure due to the pressure gradient force. The coriolis force immediately deflects this perpendicular to its present path.

The second effect on the horizontal air flow occurs because the Earth rotates, producing a force called "geostrophic" (earth-turning) or *coriolis* (after the French mathematician Coriolis who first described it). The effect of this is to appear to deflect the air-flow from going down the pressure gradient in favor of going along the gradient. In the northern hemisphere the apparent deflection is to the right, and in the southern hemisphere to the left. The combined result of the pressure gradient force and the geostrophic force is to make the air flow parallel to the lines of equal pressure (isobars), with low pressure on the left and high pressure on the right in the northern hemisphere, and high pressure on the left and low pressure on the right in the southern hemisphere. So if a mass of air starts to move southward down a pressure gradient in the northern hemisphere it will move directly north to south with reference to space; but as the Earth rotates eastward, the air gradually assumes a northeast to southwest movement relative to the Earth's parallels and meridians.

This flow, known as the *geostrophic wind*, is very nearly equal in speed and direction to real winds. However, the geostrophic balance is true only for straight isobars. In reality the pressure distribution most often takes the form of curved isobars, and this results in a third force, the *centripetal* (center-seeking) force, that influences the flow by pushing it inward. This centripetal effect occurs because the pressure gradient force is slightly stronger than the geostrophic force. The diagram illustrates the balanced flow around low and high pressure areas. This is known as the gradient wind. For the same pressure gradient, the wind speeds are higher

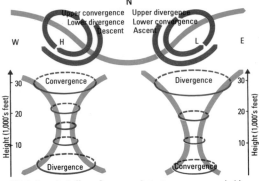

High-level airflow in a Rossby wave (top). On the upwind (W) side, the airflow at high levels *converges*; on the downwind (E) side it *diverges*.
At low levels (bottom), anticyclones occur with upper-level convergence; cyclones with upper-level divergence.

Jet streams are probably due to strong horizontal gradients in temperature which induce the thermal wind mechanism. The high-level cloud seen here is probably associated with a jet stream and consists of a cloud band about 60 miles wide and several hundreds of miles long.

round a high pressure center than around a low pressure area.

The final major effect on winds is due to *friction*. Near the Earth's surface the frictional drag on the air reduces its speed. Now the wind blows across the isobars, by as much as 15° toward the center of low pressure. At the Earth's actual surface, the air does not move, whereas about 3,000ft (1km) above the surface, the frictional effect is lost and the winds are at gradient value. Throughout this layer (the boundary layer) wind speed increases with height, as does the geostrophic nature of wind direction.

The four forces described combine in several ways to produce a multitude of airflows, causing the atmosphere to move in large-scale wind belts over the Earth. These belts are the *doldrums*, the *trades*, the *prevailing westerlies* and the *polar easterlies*.

Above the boundary layer, winds increase in speed with height, with localized (on a large scale) cores of high-speed flow. These cores emerge on global maps of airflow as very similar to meandering rivers. They are called *jet streams*, tend to be a few thousand miles long (in meridional or longitudinal direction) and about 7,000–10,000ft (2–3km) deep. They lie at heights of about 33,000–50,000ft (10–15km) (depending on latitude) at breaks in the tropopause and are frequently associated with a strong, horizontal temperature gradient (a front, see p. 26). This front is the key to their origin.

The fall in air pressure with height is less in warm air than in cold air. So although surface or low-level pressure may be equal at both a warm and a cold site, there will be differences in pressure as height increases. The air flows under the influence of both the pressure gradient and coriolis forces. Even though there may be no

wind at ground level, this temperature gradient has caused a wind
at higher levels. The process is known as *thermal wind mechanism* and
frequently simply as the thermal wind. The general rule is that the
upper wind is equal to the addition of the thermal wind to any
wind that exists lower down.

The thermal wind mechanism is one of the most important in
meteorology. It goes far to explain the jet streams, which lie astride
strong horizontal gradients of temperature. Additionally it under-
pins the mechanism of the familiar sea and land breezes where the
horizontal temperature gradient results from the land-sea contrast.

Fluid flows are complex and the atmosphere well exemplifies
this. To clarify matters, complicated flows are broken down into
simpler types. Four such types are *translation, deformation, conver-
gence/divergence*, and *vorticity* (spin). Any flow can be described by
adding together appropriate amounts of these basic flows. Translation means uniform flow in a straight line; deformation can
create gradients, such as those found in the distribution of temperature at fronts; convergence/divergence can do exactly what their
names suggest, converge and diverge, but they also commonly
occur in a straight flow. When the air speed decreases downwind,
then the flow converges. It is rather like the convergence of
vehicles at the end of a freeway, resulting from the slowing of
those nearest the exit. Conversely if air speed increases downwind,
it diverges like faster cars speeding away from traffic lights and
thus reducing the concentration of vehicles behind the lights. Spin
also has dual character. Spin in the same sense as the planet is
cyclonic and deemed to be positive. This is counterclockwise in the
northern and clockwise in the southern hemisphere. Spin in the

opposite sense is anticyclonic and negative.

These notions of spin and divergence and convergence are very important to the description and analysis of weather systems outlined previously. So air motion is fundamental to weather and climate. It is slightly ironic that one of meteorology's most powerful notions, the geostrophic wind, would, if it existed everywhere, prevent the formation of most of our weather. This is because the winds flow along the gradients and thus inhibit vertical motion, which is vital to the formation of cloud and precipitation. Fortunately, the atmosphere is not totally geostrophic: in fact it is just sufficiently non-geostrophic to allow the significant growth of weather systems.

Vertical motion

Vertical motion is of primary importance in meteorology because most weather and climate are closely and directly related to it. For example, rising air is the primary cause of cooling, condensation, cloud and precipitation. In turn the cloud shields the ground and reflects sunlight, with an effect on radiation and temperatures. Conversely, sinking air inhibits the growth of cloud and precipitation, thus allowing free play of both solar and terrestrial radiation which in turn influences temperatures. Thus, most of the main weather elements that impinge directly on us – temperature, humidity, cloud and precipitation – are strongly influenced by vertical motion.

Vertical motion in the atmosphere occurs on both large and small scales. Large-scale vertical motion occurring over thousands of square miles is found in mid-latitude cyclones and anticyclones, and has speeds of a few inches per second. The main cause is the distribution of the divergence/convergence of horizontal winds. If the horizontal winds converge into a volume then, because the atmosphere is locally incompressible, air must move vertically to retain a constant density. If the bottom of the volume is the ground, all the movement will be upward. At levels just below the tropopause convergence of the largely horizontal flow leads to sinking air. The air cannot rise because the tropopause acts like a lid. When the sinking air reaches the ground it must, of course, diverge. Rising air results from low-level convergence overlain by high-level divergence. Between the two altitudinal extremes lies a level at which no divergence/convergence exists – the level of non-divergence.

Pressure decreases with height and so a pressure gradient force acts upward. In general this force is balanced by gravity. If the two do not quite balance, the difference is called a *buoyancy force*.

Small-scale vertical motion occurs over a few square miles at most and may reach speeds of several feet per second. This is due to the buoyancy forces. If the atmosphere is such that a volume of air gathers speed as it rises or sinks, instability exists. Conversely if, after being forced away from its original height, the volume of air returns to that height, then the atmosphere is stable. The presence or otherwise of these stabilities can be determined by examining the vertical distribution of temperature, because this relates closely

Vertical motion is arguably the most important direct cause of weather. In sinking air, air is warmed adiabatically and cloud growth is inhibited. Conversely, in rising air, clouds may form. These result from the cooling of air due to adiabatic expansion and the condensation of vapor.

The uplift has two main forms at the local scale; unstable, bubbling lift as manifest in cumulus clouds, and stable, layered lift as manifest in stratiform and some other types of cloud. The picture shows clouds resulting from lift in a stable atmosphere.

to buoyancy. When a small volume of air moves vertically it changes temperature at a known rate. If the rate of temperature change in a vertically moving parcel of air is compared with the rate in the surrounding air, it is immediately apparent whether the parcel is warmer, cooler or the same temperature as the environment. In turn this identifies the buoyancy or lack of it and hence the vertical motion.

In addition to the above purely atmospheric processes, the Earth's hills and valleys have an important effect on vertical motion. Clearly air rises when it is forced to move over rising terrain, which may be abrupt as in the English Lake District or as gentle as the US Great Plains. Additionally, bubbles of air may become unstable over thermal hot spots such as plowed fields, cooling towers or urban areas. These are the mechanical and thermal effects respectively of the Earth's topography upon the atmosphere.

Some extremes of pressure and wind

Pressure is easier to record accurately than is airflow, so it forms the foundation of the weather map. It also has physiological effects on the human body, causing ears to "pop" at higher altitudes where the pressure is less.

Very high pressures tend to be found over the northern continents in winter. In December 1968 a sea level pressure of 1083.8 millibars – over 70mb higher than average – was recorded at Agata, northern Siberia. Throughout the area the skies were practically clear with temperatures between $-40°F$ and $-58°F$ ($-40°C$ and $-50°C$), with no wind and occasional mist and fog in some localities. These are the classic conditions. The lack of wind resulted from the lack of significant pressure gradients, a basic feature of high pressure, anticyclonic areas. If such conditions occur over industrial areas, they strongly favor the accumulation of pollutants.

In contrast to high pressures, low pressures are associated with winds; usually the lower the pressure, the stronger the winds. Again, this occurs because the lower the pressure in one area, the stronger the gradient between that area and another with pressure values nearer the average. Many different situations occur but three types serve as illustration: tornadoes, tropical cyclones and extratropical cyclones.

Wind speeds in tornadoes cause many problems, on occasion leading to widespread damage and death. Because of this it is very difficult to get good measurements of wind speed – the instruments are destroyed by the thing they are trying to measure. Tornadoes occur frequently over the central states of the USA and their ominous sight, sound and devastating effects are all too familiar to natives of those areas.

The world's records, as opposed to estimates, of extreme surface low pressures have been in tropical cyclones. Two measurements of 877mb have been made – one in typhoon Ida in 1958 and one in typhoon Nora in 1973, both in the western Pacific. Wind speeds of over 170mph (273kph) were recorded in Nora. The eye was about

High wind speeds result from strong horizontal pressure gradients. The strongest occur in tornadoes where winds of over 200mph may strike, destroying virtually everything in their path. Tropical cyclones have less intense pressure gradients but they, and hence the associated high winds, cover far larger areas.

9 miles (15km) in height. Although the core surface pressure rose to 894mb ten hours later, and to 924mb a day later, these are still very low values and Nora remained a very severe storm for several days. This contrasts with the tornado, which lasts for hours rather than days.

Very low pressures and high wind speeds are less frequent in extratropical cyclones. Nevertheless, they cause, at worst, death and destruction and at best gross inconvenience for the inhabitants of the mid-latitudes. In December 1982 a cyclone with a central surface low pressure of 931mb passed over the northern British Isles. This value is a twentieth century record for that part of the world and is representative of the deepest extratropical cyclones. On December 8 1886 pressure fell to 927.2mb in Belfast. This very low central pressure caused a pressure gradient of about 50mb over a distance of about 930 miles (1,500km) and severe gales ensued with gusts over 90mph (145kph).

High wind speeds at the surface are more frequent in mountainous areas, largely because the frictional drag on summits is far smaller than on lower ground. This is well exemplified by the speed of 199.5mph (321kph) recorded at Cannon Mountain (at more than 5,000ft/1,524km), New Hampshire on April 2 1973. Even the average hourly speed on this occasion touched 140 mph (225kph).

At lower levels in mountainous areas, the generation of lee waves may result in very high speeds at the ground at relatively low altitudes. Thus at Boulder, Colorado, which, although at about 5,000ft (1,524m) above sea level, is about 7,000ft (2,133m) below the peaks of the highest Rocky Mountains, strong lee wave development has given gusts of over 125mph (200kph) resulting in insurance claims over a million dollars.

The base of a tornado (top). The funnel is a few tens of feet wide in general, but at the base it appears wider because of the disturbance of the surface and the influx of material. Uplift can be particularly strong, capable of lifting large farm animals, and horizontal winds have driven stalks of straw through window panes.

Tornadoes form in cumulonimbus clouds which are severe convective storms (left). The funnels originate in the rear right quadrant of such a storm where cyclonic spin develops. This spin concentrates and the funnel develops from the base of the cloud down to the ground. The tornadoes move at 25–40mph and may occasionally skip within their tracks, alternately leaving and returning to the surface.

The water balance

Water on planet Earth is constantly recycled. Evaporation and transpiration transfer water from the Earth's surface to the atmosphere in vapor form. Once there it may be blown about as vapor or it may be turned into liquid or ice after cooling due to uplift. This gives us the familiar clouds. In turn, the clouds may produce precipitation (rain, hail, snow) which returns the water to the Earth. At the surface, water infiltrates and runs off, replenishing the store for further evaporation. Within this cycle the atmosphere has less than 1% of the water mass, the greater bulk being in the oceans.

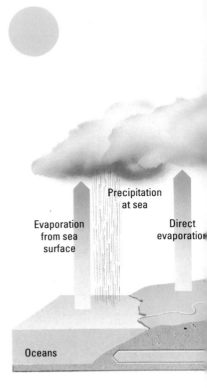

Precipitation at sea

Evaporation from sea surface

Direct evaporation

Oceans

We have seen how important heat energy is to atmospheric behavior and the creation of weather systems; it is also vital for life. Water has a similar dual role, being essential for life and also playing a key part in the general circulation of the atmosphere. The water molecule is the heat carrier of the atmosphere, storing energy from the Sun and releasing it in the process that we call the *planetary hydrological cycle*.

Water circulates through different parts of the natural environment in many ways. Ice and liquid water occur in glaciers, rivers, and lakes, and especially in oceans, which contain over 97% of the water on Earth. The part played by the atmosphere is to receive water, taking it from the water surface by means of evaporation and *transpiration*: to move it long distances as water vapor and as cloud droplets; and to return the water to the ground by precipitation – rain, snow, sleet hail etc.

Evaporation

Evaporation is the process whereby water turns from liquid to vapor. At any water surface, water molecules are continually leaving and returning to the surface. When more leave than return, evaporation occurs; when more return than leave, condensation occurs. The water molecules require energy (heat) to leave the surface, and this heat is taken from the immediate surroundings. This heat that is removed from the immediate sur-

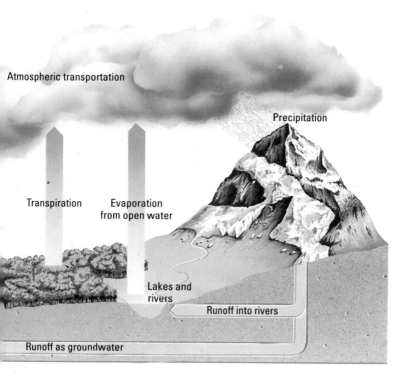

Atmospheric transportation

Precipitation

Transpiration

Evaporation
from open water

Lakes and
rivers

Runoff into rivers

Runoff as groundwater

roundings is called latent heat, in contrast to the "sensible" or direct heat that we feel. When the molecules return to the surface they release their latent heat, as when condensation occurs on a cold pane of glass. The extraction and release of latent heat leads to the cooling and warming of the air in which it occurs.

The warmer the air, the more water vapor it can hold; the cooler the air, the less water vapor it can hold. The rate of water loss by evaporation and transpiration (the process whereby plants lose water to the air) depends on the supply of energy or heat at the surface, and also on the difference in the saturation vapor pressure at the surface and the vapor pressure of the air. Saturation is reached when the air's capacity to hold more water vapor is zero and the number of molecules returning to the water surface is equal to the number leaving. Beyond this point, the water vapor becomes visible, as in a cloud or fog.

The largest values of annual average evaporation occur over the subtropical oceans, where there is no shortage of water and the amounts of solar radiation are, as we have seen, as high as anywhere on the planet's surface. Here the evaporation averages about 79in (200cm). Over the oceans in the higher latitudes the annual amount falls to about 20in (50cm) or less. This reflects the lesser supply of heat energy in these regions. Evaporation over land areas may be as low as 11in (30cm) over the hot dry deserts, and as high as 40in (100cm) over the tropical rain forests.

Humidity

In the atmosphere, water occurs in all its forms: solid, liquid and gaseous. Humidity is a word used to describe water in its gaseous phase, or the amount of water vapor in the air. This moisture content in the air can be measured by means of a *psychrometer* consisting of two thermometers attached to the same frame – the dry-bulb and the wet-bulb thermometer. The first is an ordinary, unmodified thermometer. The second is an ordinary thermometer with wet muslin around the bulb. As the water in the muslin evaporates it draws heat from the air and so the temperature falls. The drier the air, the more evaporation takes place and the lower the wet-bulb temperature becomes. So the greater the difference between the dry-bulb and the wet-bulb temperatures, the lower the humidity.

Everyday experience of humidity tends to manifest itself in discomfort. When the relative humidity – the ratio between the amount of water vapor in the air and the saturation level of the air at the same temperature – is high, the result can be uncomfortable in both hot and cold air. In hot air, a high relative humidity results in "muggy" conditions; the body does not easily lose heat through sweating because the moist air hinders the evaporation of sweat. Hence the uncomfortable hot summer days so frequently experienced in the eastern United States.

If the temperature is lower, though the quantity of water vapor is the same the relative humidity will be higher. So a

Welcome relief from a fire hydrant in New York where, in summer, the combination of high temperatures and humidities results in an oppressively muggy atmosphere. Air conditioning becomes a necessity in such conditions.

relative humidity of 60% will feel far more "muggy" at 75°C than at 45°C. High humidities in cold air give it a "raw" quality, which explains why British winters, while being remarkably mild for their latitude, may seem more unpleasant than the very cold but drier air of, for instance, the Midwest of the United States.

Very low humidities are less obviously felt, but occur more frequently than might be expected. As the dry air associated with anticyclones descends it gets warmer, which reduces the relative humidity. The same mechanism operates in the downslope winds associated with lee waves on mountains. These winds, known as Föhns (originally the wind that blows down the northern slopes of the Alps), are notable for their warmth and dryness.

The atmosphere receives water from the Earth's surface, so it is hardly surprising that the highest humidity values are near the surface. In fact, the bulk of the atmosphere's water lies below a height of about 18,000ft (5.5km). Another way of presenting humidity is to express it as "precipitable water," or the depth of liquid water that would result over a given area if all the water vapor in the column above that area were turned into liquid. Averaged over the whole world, the depth of water contained in the atmosphere is only 1in (2.5cm). Locally, in the Asian summer monsoon, the amount rises to 2.4in (6cm). Thus the average amount of water put into the atmosphere in a year by evaporation is, as we have seen, about 100 times what the atmosphere, on average, holds. The difference is accounted for by precipitation.

Dew, fog and smog

The cooler the air, the less water vapor it can hold. Cooling of air eventually leads to saturation and the water vapor condenses into droplets. These droplets may be deposited on the Earth's surface in the form of either dew or frost (depending on the temperature), as often happens after nighttime cooling.

Water vapor also condenses onto extremely small nuclei in air itself – smoke, seasalt, sulfate aerosols and sulfur dioxide, for example – to form very small droplets of water, droplets small enough to remain suspended in the air rather than fall to the ground. These account for the clouds we see or, if they blanket the ground, they are known as mist or fog (depending on their concentration).

In areas where quantities of industrial smoke rise into the atmosphere, water vapor may condense onto airborne chemical wastes. The resulting mixture of smoke and fog – smog – is particularly dangerous for the elderly, people with heart conditions, asthmatics and the physically weak. Many authorities insist on smoke control procedures, and in Los Angeles – an area particularly at risk – special mufflers have been designed for automobile exhausts to cut down the contaminating effects of fumes.

A satellite picture of San Francisco Bay shows thick banks of fog forming over the Pacific Ocean and approaching the warmer land area of San Francisco. The clarity of the Bay and the apparent channeling of the fog through the Golden Gate Bridge (see opposite picture) is strikingly evident, emphasizing how this type of fog forms through cooling of air over the colder sea.

The Golden Gate Bridge is seen to be shrouded in advection fog — a type of fog formation in which the ocean surface cools the overlying air to condensation point, whence the fog forms. As it moves inland, the greater warmth of the land surface may bring about a rapid evaporation of the fog, causing it to break up and thin out as it is carried over the ''heat island'' of urban San Francisco.

Radiation at night leads to cooling of air; should this persist in air with sufficient moisture, then condensation frequently occurs and fog forms. If this process goes on in irregular terrain, then the cool air flows down the slopes, frequently accumulating in valley bottoms. If the air is foggy, then the valley may accumulate the fog, as shown here, giving the impression of a lake. Once formed, the fog encourages its own growth by radiational cooling from its upper surface.

Valley fogs frequently have smooth tops, rather like a pond. On other occasions far more structure is visible, as here, reflecting either the turbulent motions within the fog or, more rarely, the growth of waves on the surface of the fog.

Fog types

Cooling fogs and *evaporation fogs* represent the two basic kinds. The first includes two main methods of cooling – radiation and advection. In the first case, when radiation from the ground lowers surface temperature, this cooling is in turn transmitted to the overlying air. A very gentle turbulence in the air helps to distribute the cooling over depths of tens of feet – otherwise it would remain in a very shallow layer on the surface. Once the cooling has reached the *dew point* temperature, that is, the temperature at which saturation occurs at constant pressure, the water vapor condenses and fog droplets form. The fog top itself then becomes the radiating surface in turn, leading to further cooling and a probable deepening of the fog. This mechanism operates most efficiently in long nights with clear skies, such as frequently happen in the fall in mid-latitudes. In hilly areas, air flowing downhill would aid the process, leading to the accumulation of fog in the valleys – a very widespread occurrence.

The second type of cooling fog results from advection (see p. 52). Air masses that move away from their areas of origin to cooler surfaces experience a progressive cooling on their lowest layers. If the surface also happens to be water, the moisture content of the air will, at most, be increased or, at least, maintained. The combined effects of cooling and increased vapor content clearly favor fog formation. The classic areas for these ''advection fogs'' are the Newfoundland Banks, where air from the warm south passes over the cold Labrador current, and in the southwestern areas of the British Isles. In contrast to the radiation fogs, the advection fogs are not dependent on time of day, and may persist for tens of hours. They dissipate when the large-scale airflow

Shallow fog and low cloud in mountain and lake country. The fog is formed over the water and lifted slightly as the air meets the surrounding mountains. Advection fogs are caused by the transport of warm moist air over colder surfaces.

A shallow, wispy evaporation fog, caused by the same mechanism as arctic smoke. Vapor from the relatively warm water surface is immediately condensed in the overlying cold air — as in the formation of steam from boiling water.

Frost occurs when the temperature of the Earth's surface falls below freezing. It takes a variety of forms, the most frequent being hoar (or white) frost. The freezing temperatures ensure that water on the surface becomes ice.

changes, usually to a form which leads to the mixing of foggy and clear air: this mixing leads to evaporation of the droplets in the resultant non-saturated air. Radiation fogs dissipate by two processes: the sun may burn them off; or the wind may increase vertical mixing with the same result as in advection fog. In many cases, of course, the winter sun is inadequate to burn off the previous night's fog and so it continues, and probably deepens, through the next night. Ultimately it is then usually the wind which gets rid of the fog.

Evaporation fogs are a little more complicated than cooling fogs. They result from the immediate condensation of water vapor that has just evaporated from the underlying surface. The surfaces thus appear to "steam." One can usually see this after a heavy shower and the most widespread occurrence is over relatively warm water in very cold air. Hence the so-called *arctic smoke*, which is found at high latitudes over the relatively warm water of the North Atlantic and Pacific Oceans.

Cloud formation

Clouds are made out of the same "material" as fog – water droplets or ice crystals that have formed around microscopic nuclei in the atmosphere. A number of processes are responsible for the formation of clouds and their ensuing shapes and sizes.

Clouds, like fog, are caused by the cooling of air so that water condenses. But whereas fog forms from cooling to dew point temperature at a constant pressure, clouds form due to the uplift and expansion of air. The higher the altitude, the lower the pressure, so when a parcel of air rises it moves into regions of increasingly low

A shower cloud with familiar icing in its upper reaches. The cloud is a few miles across and deep and the icing sets off a mechanism to produce rainfall.

atmospheric pressure, and the air volume expands. This expansion requires energy which is taken from the heat of the air, and so the temperature falls. This is known as *adiabatic cooling*.

Condensation and freezing occur around appropriate nuclei – processes which result from adiabatic cooling, which in turn results from uprising air. But what is the nature of this uplift?

Two main types of uplift occur: slow and widespread, and rapid and localized. The former is a feature of extratropical cyclones, particularly at a warm front – the advancing edge of a mass of warm air that is displacing a mass of colder air. When moving at a speed of a few inches per second, the air takes above five hours to rise half a mile, so this kind of uplift may last many hours to allow the observed persistence of clouds within cyclones. Within such widespread, steady ascent the condensation rates are also steady and the resultant droplets are rather small. The ensuing cloud will be like a sheet – uniform, covering large areas, and being about half a mile deep. This is called "stratiform" cloud (the Latin word *stratum* means a layer or sheet). It can occur at most levels in the lower part of the atmosphere but is most frequent below about 16,500ft (5km). This kind of cloud is ubiquitous over the globe and is largely responsible for the average 50% coverage of the Earth – as dramatically seen on satellite photographs.

Localized uplift results from convection. Thermals a few hundred feet across bubble up to the condensation level, at which point we see them as cloud. The resultant "heaped" shape is called "cumuliform" (the Latin word *cumulus* means a mass). In contrast to the stratiform clouds, cumuliform clouds tend to have distinctive outlines, to be a few miles across, frequently to be as deep as they are wide, and to exist for a few hours. Occasionally they may become as deep as the troposphere (the lowest layer of the

Well-developed cumulus congestus clouds. Cumuliform clouds result from thermal convection lifting heat from the surface. After adiabatic cooling the vapor in the air condenses to give the familiar cauliflower outline, with flat base and sharp, rounded top. If the atmosphere is particularly unstable the clouds may continue to grow in the congestus form shown here, where the individual towers are about twice as high as they are wide. As the clouds grow they draw in surrounding clear air. This mixes with the cloudy air, leading to evaporation of cloud droplets and hence erosion of the cloud. As a result cumulus congestus tend to be short-lived, either developing further into a larger cumulonimbus cloud or dying away.

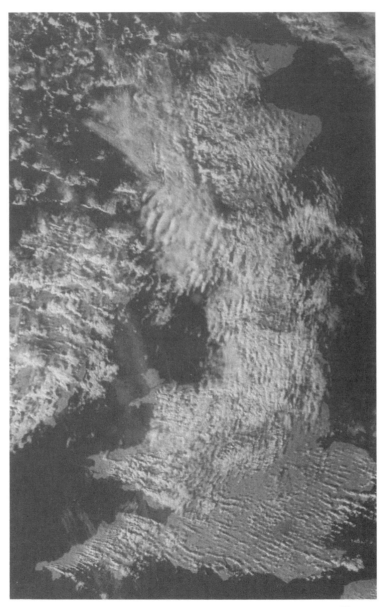

A satellite picture of clouds over the British Isles shows three main patterns. First, in the northwest, shallow, cellular convection occurs, with a hint of honeycomb structure. Second, over Scotland and northern England are trains of lenticular clouds in the crests of lee waves. These crests (and clouds) are about 6 miles apart and the trains extend for well over 60 miles. Third, over southern England, well developed cloud streets run from roughly west to east. Note that the orientations of the lenticular and cumulus clouds are virtually perpendicular to one another – a striking illustration of different mechanisms occurring in the same airstream.

atmosphere). This type of cloud is associated with instability in the atmosphere (see pp. 35–36): in general terms the greater the instability the more vigorous and the deeper the cumuliform cloud.

Mountains and valleys are an important cause of uplift. They affect both basic types of cloud, tending to intensify an already existing stratiform cloud and to encourage growth of cumuliform cloud over hilly areas at the expense of the surrounding lowlands.

Cloud classification

The apparent simplicity of two basic mechanisms of cloud formation hides myriad actual cloud shapes, sizes and colors. The tremendous variety of the sky has long been appreciated and it led to an attempt to find order over 150 years ago. Luke Howard, an English pharmacist, proposed a classification which, in its essentials, remains valid today. The system is based upon two descriptive criteria – shape and height – rather than on how clouds are formed. Whilst this approach has been criticized, and alternative schemes proposed, it is still used because it is simple, requiring only careful observation and not a sound knowledge of atmospheric processes. Howard recognized three height categories: high (20,000ft/6km or above); medium (6,500ft to 20,000ft/2km to 6km); and low (below 6,500ft/2km). Using these criteria, ten basic cloud types were identified, eight of them falling into the three

Mountain clouds form over the Himalayas. As air is forced over mountains, the vapor within it frequently condenses into cloud. Mountain ranges block the free movement of air masses, sheltering areas on their leeward sides. Thus the Himalayas prevent the cold continental air masses of north and central Asia, which make east Asian winters so cold, from entering the sub-continent.

height categories and two occurring in more than one height range.

At the highest levels three types occur: cirrus, cirrostratus, and cirrocumulus. These are ice clouds having a fibrous or feathery appearance. Cirrus are detached clouds comprising white, delicate filaments. Cirrocumulus is a thin layer of cloud without any shading, comprising very small elements in the form of grains or ripples, merged or separate and fairly regularly arranged. Cirrostratus is a transparent, whitish cloud veil of fibrous or smooth appearance over most of the sky.

At the middle levels two types occur: altocumulus, and altostratus. Altocumulus tends to be a sheet of small, rounded masses, frequently merged together. It is commonly called a "mackerel sky" because of the similarity of its pattern to that on the fish. It is frequently the precursor of rain. Hence the saying: Mackerel sky, not long dry. A particularly striking form of altocumulus is the subtype called "lenticularis." These are the lens-shaped clouds associated with waves in the atmosphere (see p. 30). Altostratus is a gray sheet of cloud with a fairly uniform appearance. As the name suggests, it is a stratiform cloud at middle levels.

At low levels three types occur: nimbostratus, stratocumulus and stratus. Nimbostratus is a gray cloud layer that is precipitating (*nimbus* is Latin for "rain"). It is therefore often dark and diffuse with low, ragged clouds below its main base.

Well-developed bank of cumulus cloud hides the sun and causes crepuscular rays to be visible on the right of the picture. The rays are alternatively lighter and darker bands which appear to diverge in a fan-like array from the sun's position. The divergence is due to perspective. The back-lighting of the clouds outlines the heaped nature of the cumulus clouds.

Cirrus clouds lie at high levels (6 miles) and hence are made of ice crystals. They are of three main types: cirrostratus, cirrocumulus and cirrus. The last can cause some dramatic, disturbed skies as shown here, where the air seems to be streaming out of the picture from behind the silhouetted mountain.

Satellite view (right) of shallow cellular convection. Light areas are cloud, resulting from uplift at the mesoscale. Dark areas are clear air where the compensatory subsidence occurs. The cloudy areas are several miles wide. This type of convection occurs in cold air over warm oceans such as is found when cold Siberian air flows over the Sea of Japan. It is a relatively recently observed feature and no full explanation has yet emerged.

A storm cloud (above) approaches Capocabana Beach of Rio de Janeiro, yet the sun still plays on the sea through a break in the clouds. Tropical cumulus can give very intense rainfall.

A sky full of waves (below). Lenticular clouds formed in the crest of waves over the Cairngorm Mountains of Scotland. The airstream was fairly stable just above the mountain tops and hence, in moving over the summits, vertical oscillations were set up. As the air flowed from west to east (out of the picture), waves were formed and hence this dramatic cloudscape. The clouds are stationary relative to the Earth despite the rapid movement of air through them.

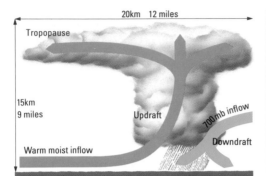

20km   12 miles

Tropopause

15km
9 miles

Updraft

700 mb inflow

Downdraft

Warm moist inflow

Cross section of airflow within a well-developed cumulonimbus cloud (above) compared with an isolated natural formation (right). The cloud has the distinctive anvil shape and moves from right to left. Warm, moist air enters at the front, low levels and rises towards the rear of the storm. Behind and underneath the updraught lies the cool, dry downdraught which originates in middle levels. The cool air splays out as it hits the ground, the forward-moving part helping to create new uplift at the storm's leading edge.

A typical fair-weather cumulus sky. Thermal convection leads to uplift of moist air, which, at its lifting condensation level, results in cloud. Frequently the vertical temperature and humidity structure of the atmosphere inhibits further growth, resulting in shallow clouds.

Stratocumulus is a cloud that has rounded masses within it, reminiscent of cumuliform, hence the juxtaposition of the two basic terms. Stratus is the ubiquitous gray cloud layer with a uniform base. If it precipitates it becomes nimbostratus.

The two remaining cloud types that cut across the height categories are cumulus and cumulonimbus. Cumulus are the very familiar heaped, cauliflower-like clouds, with sharp outlines and fairly flat bases. In sunlight the upper parts are brilliantly white. Cumulonimbus is the deepest and most vigorous of clouds. It may extend throughout the depth of the troposphere and, if so, will certainly comprise ice crystals in the upper reaches. These give the cloud a smooth, fibrous outline which is frequently seen in a distinctively anvil shape, due to the spreading of the ice crystals by the high-level winds.

These ten types are but the core of a great multitude of cloud forms that actually occur in the atmosphere. Moreover, the types and their variations can coexist at any time. This leads to all sorts of interesting and complicated skyscapes that have intrigued sky watchers for centuries. Despite the satellites that give us a different perspective on clouds – viewed from above not below and seen over huge areas, not just the individual's sky – we still need the Howardian classification.

Precipitation

Water leaves the atmosphere as precipitation, a term that covers all liquid and solid water that falls from the sky and reaches the ground. Over 20 types of precipitation have been identified.

Liquid and solid precipitation begin in two microscale processes, one requiring condensation of water vapor around microscopic particles, known as condensation nuclei, the other requiring the freezing of water droplets onto freezing nuclei to give very small ice crystals. These crystals subsequently grow by direct deposition of vapor onto them until they reach a size where they may begin to collide and join with each other, producing larger entities which are snowflakes. This process, called *aggregation*, is particularly efficient at temperatures between 32°F (0°C) and 23°F (−5°C). In their early stages the ice crystals are encouraged to grow by the coexistence of water droplets with temperatures between −4°F (−20°C) and −40°F (−40°C). These *supercooled* droplets remain liquid at subfreezing temperatures because there is no microsopic ice or freezing nuclei in the atmosphere for them to freeze around. One of the properties of water is that at temperatures less than 32°F (0°C), where both ice and liquid exist, the saturation vapor pressure over water is greater than that over ice. Vapor that is just saturated with respect to the supercooled water droplets will be *supersaturated* with respect to the ice. This results in vapor being deposited onto the ice crystal. in turn, the reduction

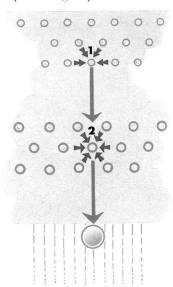 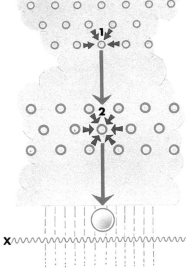

Raindrops are formed when the microscopic droplets of water vapor (1) condense onto airborne nuclei such as tiny particles of sea salt. There they merge to form larger droplets (2) which will eventually become too heavy to be kept up by air currents.

Water can remain liquid at well below freezing point in still air. As such it is termed "supercooled", and occurs in large quantities above the freezing level (X). Should it fall as rain, supercooled water forms an icy glaze on any object it encounters.

in vapor means that it is no longer saturated with respect to liquid water and so the water droplets begin to evaporate in an attempt to achieve saturation once more. This process means that the ice crystals grow at the expense of the water droplets, in turn encouraging formation of snowflakes which may reach the ground or, as most frequently happens, melt at the 32°F (0°C) level and become raindrops. This whole process (known as the Bergeron-Findeisen theory) explains most extratropical snow and rainfall.

Some clouds do not contain a mixture of ice and supercooled water: in them, the particles grow by collision and coalescence. Ice to ice is called aggregation; ice to water is called *accretion*; and water to water is called *coalescence*. It is the last process which probably accounts for the precipitation from clouds, usually in tropical latitudes that contain no ice. Accretion is the major process whereby hailstones are formed, an embryo of ice collecting small liquid droplets which freeze on impact with it.

All these small-scale processes operate over periods measured in minutes and hours within the myriad clouds throughout the atmosphere. The amounts and types of precipitation found at the ground are the end products of a chain of processes ranging in scale from inches to thousands of miles, because these very small-scale mechanisms operate within clouds which themselves lie within weather systems. Hence the global pattern of precipitation reflects the distribution of such systems.

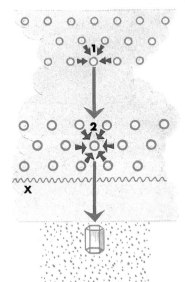

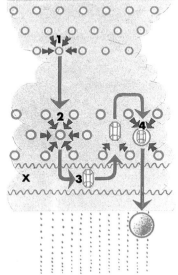

Snow formation is quite different from that of rain. The tiny water droplets (1) settle and freeze onto microscopic ice crystals which thereby increase in size. The enlarged crystals then form characteristic clusters which fall to the ground as snowflakes.

Hailstones occur when raindrops formed in cumulonimbus clouds (1, 2) are subjected to strong up-currents. Thus they may be taken back through the freezing layer (X) several times, collecting a fresh layer of ice (3) until weight finally forces them downwards.

The world's wettest mountain – Waiaileale in Hawaii. Warm, moist air is constantly forced over this mountain by the trade winds, giving a semi-permanent cap of cloud with very high water content.

Rainfall patterns

The amount of precipitation of any type, rain or snow, that the ground receives, is related to the speed with which the weather system releases the precipitation, and the speed of the weather system over the ground. Obviously, an area beneath a stationary system will receive more precipitation at a given release rate than an area passed over by a moving system of the same type. So the largest amounts of rain, snow or hail come from slow-moving or stationary cloud and weather systems. The rate of release is measured indirectly at the ground as the intensity of precipitation – the amount received over a specified period. Generally, very large intensities – more than 4in (100mm) per hour – occur for short periods of minutes, whereas lower intensities – 0.04in (1mm) or less per hour – continue for tens of hours. The rates and durations are related to the vertical speeds and durations of the clouds that spawn them: intense, shortlived falls from vigorous cumuliform clouds, and gentler but longer-lasting falls from stratiform clouds.

The heaviest falls of both rainfall and hail come from cumulonimbus clouds. In the tropics these are capable of depositing on one place 72in (1800mm) of rainfall in a single day. Even in the UK, a particularly vigorous storm gave more than 7in (170mm) in about three hours, and this broke all national records for sub-daily amounts and intensities. Thunderstorms give the heaviest rainfalls.

Prolonged rainfall also occurs in the warm sectors of cyclones, and because of mountainous terrain, particularly if these two

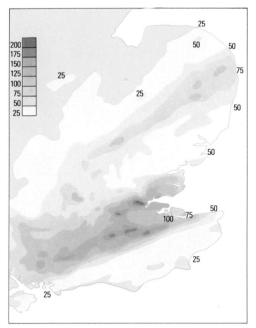

Map of rainfall from a severe rainstorm over southeast England. Note the amounts, the linear form of the distribution and the abrupt southern edge. These heavy falls came from thunderstorms that formed on an occluded front (blobs and spikes) lying over southeast England (bottom map). The lines are isobars and the shading is the rain area. A low pressure center lies over the region receiving the heaviest rainfall. The relatively slow movement of the precipitating systems meant that rainfall at the ground was particularly intense.

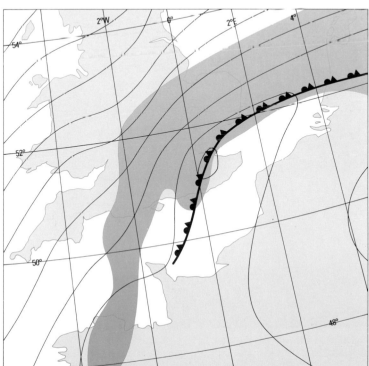

Snowfall at the equator may seem paradoxical but can occur at sufficiently high altitudes. The photograph shows Mount Kenya with a light dusting of snow as part of a typical montane climate. Note the desert vegetation in the foreground.

influences combine. The effect due to terrain is clearly evident on any map of average monthly or annual precipitation. The "enhancement" of amounts over hills is due to the raindrops from higher level cloud (the seeder) collecting and "washing out" small droplets in a cap cloud formed by mountains (the feeder). This has become known as the seeder-feeder mechanism. The largest increases occur when the pre-existing rainfall rates are high and when strong, very moist low-level flow exists. In many cases of heavy rainfall flooding ensues and flash floods such as occur on the eastern flanks of the Rocky Mountains can be particularly devastating.

Snow is the main form of precipitation at high latitudes but it can fall at the equator if the altitude is sufficiently high, as on Mount Kenya. Up to latitude 70°, and maybe even farther pole-ward, precipitation can be as rain in summer. Conversely, even at low altitudes, snow may fall in winter as far south as, for example, Florida. In the inhabited parts of North America some of the worst snowfalls come in the winter blizzards experienced in the northern Great Plains and the northeastern metropolitan states. These blizzards occur several times a year and occasionally are very severe. March 1966 provided two such: one in North Dakota early in the month and later one in Iowa and Minnesota. In both cases the severe conditions were associated with a frontal cyclone.

Between March 2 and March 5 1966, a minimum of 5in (125mm) of snow fell over an area extending from western Montana to Minnesota and beyond. In parts of North Dakota more than 30in (725mm) of snow fell and drifts of 11 to 13ft (3 to 4m) were commonplace. Temperatures were about 14°F ($-$ 10°C) and

A ground blizzard at Waterton Lake National Park, Alberta. This area of North America experiences heavy snowfall and severe blizzards that may occur on several occasions during the year.

winds reached 100mph (160kph). Driven snow reduced visibility to zero over much of the northern Plains. At Bismarck, North Dakota, visibility fell to zero for 11 hours and was only 220 yards (201m) for a continuous period of 19 hours. In contrast to previous severe blizzards this one lasted for four days in some areas, closing schools, businesses and traffic routes and killing 18 people. The livestock loss in Nebraska and the Dakotas was estimated at $12 million.

The snowfall came from a frontal cyclone that started over western Montana. To the north lay very cold arctic air in a high pressure air mass over Canada. The cyclone moved eastward, deepened and began to form an occluded front (when a cold front overtakes a warm front). All the while the clouds at the fronts and in the cyclone centre were snowing. By March 3 the system stalled over South Dakota and continued to deepen. This provided classic conditions for large amounts of precipitation (in this case snow) and high wind speeds. The deepening would be associated with vigorously rising air, which in turn spawns active clouds within which precipitation is readily formed. The diagram shows the deep depression with the strong winds, the occluded front, and the snow area on this occasion. After the prolonged blizzard bitter cold arctic air moved south over the Dakotas in the rear part of the cyclone. Temperatures fell to below −4°F (−20°C).

In this example temperatures were so low that the snow had no chance to melt as it fell. As shown on pp. 88–89, melting does indeed happen and is a major cause of rainfall in winter in north-west Europe. In upland areas we frequently find that snow falls and lies above a certain height, below which rain or sleet occurs. Hence the terrain affects not only the amount but also the type of precipitation.

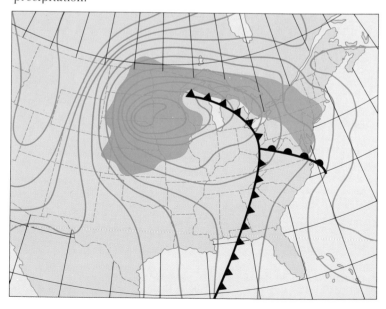

A very severe snowstorm over Dakota (opposite) is recorded in a synoptic surface weather map. A very deep, slow-moving depression gave high winds and prolonged snowfall. The closeness of the isobars shows wind speeds.

Chicago's coldest day in history occurred on January 10 1982. The temperature dipped to 26°F below zero and winds of 25mph produced a severe wind chill factor. In this fire, the water from the hoses froze almost at once.

FORECASTING THE WEATHER

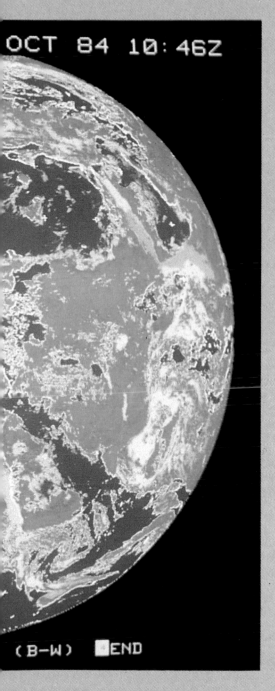

OCT 84 10:46Z

(B-W) ■END

Information from satellites is color coded to enable examination of chosen features of interest. In this image the colors are being used to emphasize regions of high reflectivity of sunlight. Deserts reflect more light than any other land, so the arid regions of North Africa and Arabia are highlighted, as well as the brighter parts of the cloud system.

Monitoring the weather machine

Understanding how weather works, and forecasting what it will be both depend on making the right measurements. And these measurements must be taken frequently and put to use with minimal delay. "Measurement" covers a wide range. It may mean a weatherman going out of doors to take readings from thermometers, wind gauges, and other instruments; or it may mean an international program to develop and launch satellites whose instruments read the atmosphere from space. Forecasters need information about the air all over the world, at surface level and at various levels above the surface. No one system of measurements can give them all the information they need; and some of the measurements are difficult to obtain by any of the available methods. Since the weather recognizes no national frontiers, there is a high degree of cooperation among the countries of the world in collecting and exchanging measurements. One of the important practical aspects of cooperation is that the same units of measurement are used by meteorologists everywhere. For distances, meters (m) and kilometers (km) are used. In the following chapters metric forms appear first with imperial equivalents in brackets.

The facts at the surface

Measurements of the properties of the air are more readily obtained near the Earth's surface, where people live, than in the layers above. Whether it be Peking or Washington DC, such measurements have more value if they are made at agreed times, and from standard types of instruments located at approved sites. The timing is important because to produce meaningful weather maps it is necessary for observations to be recorded at the same times everywhere. Such observations are called synoptic (from the Greek: "seen together"). At official meteorological stations surface observations are reported at least every three hours, and in many cases hourly.

It is also important that, as far as possible, all observations are made using standard instruments and following standard procedures. For example, if a thermometer is taken outside on a sunny day, the glass and metal parts will absorb the Sun's radiation and will be warmed to a temperature well above that of the air around it. To record the temperature of the air, which is what is needed, the thermometer must be shaded. This is why instruments are kept in the familiar louvered Stevenson screens.

Again, it is well known that wind speed increases rapidly with height through the first few tens of meters above the surface. Thus if one measurement is made at 20m (66ft) above the surface and a second at 5m (16ft) and then both values are plotted on the same weather map, some confusion will result. So an agreed standard has to be adopted, which is that measurements are made at the 10m (33ft) level. Many details of this kind, and many practical arrangements for making and exchanging observations, are discussed and agreed in the international forum provided by the World Meteorological Organization (WMO).

Even where plenty of standard synoptic observations are available, their interpretation can be troublesome. If an observing

Skilled observers in all countries record surface weather conditions every few hours. Stevenson screens, like the one on the left, shade the thermometers from the sun's direct radiation but allow the air to flow freely through.

station is downwind of well-wooded country it will record smaller wind speeds than if it were located at an exposed coastal site nearby. Or, again, at the end of a clear, still night the temperature in a city center may well be several degrees higher than in the surrounding suburbs. These local variations are of course real enough, but they are nonetheless unwanted complications when, for forecasting, the larger scale patterns have to be identified.

There are many sparsely populated land areas on the Earth. Few surface measurements are available from deserts, forests, icecaps and mountainous regions. There is an increasing tendency now to establish automatic weather stations (AWS) at key sites where it would be very costly to maintain a traditional meteorological station. The AWS are capable of transmitting radio messages containing information on pressure, temperature, wind, and humidity, but cannot mimic all the skills of the human observer, for example in identifying cloud types.

Inevitably, it is much more difficult to obtain adequate meteorological information over the sea surface. In some regions, off-shore commercial activities have led to the installation of meteorological instruments on oil rigs and drilling patterns. A number of well-instrumented moored buoys have been set up on continental shelves. Very many of the world's ships provide complete surface reports at regular intervals (usually every three hours) which are collected either by shore radio stations or by satellite. So the shipping lanes and many coastal waters are well monitored, but major parts of the oceans are left untouched, except for some very important island stations, especially in the southern hemisphere.

To help fill the gaps, drifting meteorological buoys were introduced in the late 1970s. Left to migrate in the ocean currents, these buoys have equipment capable of measuring pressure and temperature in rarely visited waters. Unfortunately though, wind measurements are difficult and expensive to obtain from such an unstable platform as a drifting buoy. To fulfill their potential, buoys need to be deployed in sufficient numbers according to a continuing and coordinated international program. The only feasible way of collecting the information they gather is to use satellites; and each satellite must also be equipped with a system for locating the buoys. Current operational satellites are capable of locating the position of buoys with an accuracy of one or two kilometers (around a mile).

Upper air balloon soundings

Weather is a three-dimensional business. Thus, even if we could know all the facts about the air at the surface at some instant of time, for the whole globe and with perfect precision, it would still be impossible to forecast what would happen next. For forecasting it is essential also to have information about the air above, and preferably to a level at least 20km (66,000ft) above the surface.

The best way of obtaining measurements of temperature, pressure and humidity through the upper air is to release buoyant, gas-filled balloons fitted with recording instruments and radio transmitters. If it were not for considerations of cost and practical difficulties, a regular, worldwide network of balloon stations would be possible – and sufficient to provide all the information forecasters need. As it is, an extensive but incomplete network has to be supplemented with other techniques.

Balloon instruments take temperature, pressure and humidity readings at frequent intervals during the balloon's ascent, and the radio transmits those readings to the ground. The information is of no use unless a forecaster knows how high the balloon was when a particular set of readings was obtained. The readings themselves are the key to that. From pressure, temperature and humidity, air density can be calculated; and density is proportional to the rate of change of pressure with height. Fairly simple mathematics gives the forecaster the answer he wants.

The other measurement needed for the upper air is wind speed and direction. Once again balloons can help. What is needed is a means of tracking a balloon's position at frequent intervals during its ascent. Then, assuming that the balloon moves horizontally with the wind as it rises, the horizontal displacement between two positions gives the average wind speed and direction for the layer through which it has passed.

A skilled observer, using a surveyor's theodolite, can take readings of bearing and elevation while keeping track of the balloon. Given a knowledge of how fast the balloon is rising, these readings are sufficient to calculate wind speed and direction. Alternatively, two theodolites can be used: the two sets of simultaneous bearings and elevations then provide sufficient information directly.

To obtain wind speed and direction through a deep layer of air, at night or in the presence of cloud, optical methods must give way to radar tracking. For this technique the balloons carry aloft a

A radiosonde system has been assembled in the special shelter and is now ready for launching. The buoyant gas-filled balloon will lift aloft a reflective target for the wind-finding radar and, at the end of the cable, a package of instruments to measure pressure, temperature and humidity and a radio transmitter to send back the measurements.

special reflective target and the radar bounces signals off this target to determine the position of the balloon throughout its ascent.

At some observing stations, particularly in the tropics, wind is the only property of the upper air to be measured, using what are termed "pilot" balloon ascents. But a full upper air report is obtained at many stations by combining the radio sounding technique for temperature, pressure, and humidity – from which height can be calculated – with the radar tracking technique for wind. These are known to meteorologists as radiosonde observations.

Beginning in the 1940s, a network of radiosonde stations has gradually been built up around the world. Efforts to improve the network still continue, but are hampered by the high cost of maintaining stations in remote areas and by other logistical difficulties. Different countries use different instruments for the same measurements on radiosondes, so there is need for calibration to a common standard.

In some parts of the world the radiosonde network is dense enough to yield an adequate description of the three-dimensional

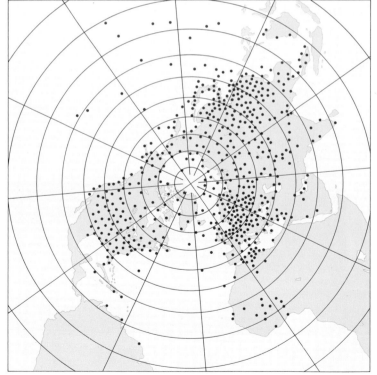

The network of northern hemisphere balloon stations from which data were available within a few hours at a forecasting center on a typical occasion. Notice the lack of coverage of the oceans and of some land areas.

structure of the atmosphere. North America, Europe and China are the regions with the best coverage. Other land areas have an incomplete but still useful coverage, whilst in certain remote areas there are just a few isolated stations. Parts of Africa are particularly poorly served.

Like surface observations, the timing of radiosonde balloon ascents is coordinated internationally so that synoptic charts of upper air patterns can be prepared. All around the world the balloons go up at midnight (0000) and midday (1200) Greenwich Mean Time (GMT) with, at a relatively few stations, intermediate ascents at 0600 and 1800GMT, sometimes limited to pilot ascents for winds. A balloon takes about an hour to rise from ground level to 20km (66,000ft) or above, so the vertical profiles are only approximately synoptic. The balloons continue to rise until the external pressure becomes so low that they burst. Specially strong balloons can be used if ascents high into the stratosphere are desired. After a balloon bursts, a small parachute slows the descent of the instrument package to prevent damage to persons or property below. The package often carries a message inviting any finder to return the instruments for recycling.

Balloons and ocean weather stations

Even at the surface the oceans are more poorly observed than most land areas. Not surprisingly, the contrast is even more marked when it comes to measuring the properties of the upper air using balloon soundings. And yet upper air information from the ocean areas is absolutely crucial for weather forecasting. It is over the oceans that the most vigorous storms develop, both in the tropics and at higher latitudes. These storms very often affect densely populated land areas later in their life cycles.

Of course, balloon soundings made from island stations are very valuable, particularly if the islands are situated in mid-ocean. The Azores and Bermuda in the North Atlantic, Ascension and St Helena in the South Atlantic, Hawaii and Fiji in the Pacific, the Seychelles and Kerguelen in the Indian Ocean, and many other islands provide key stations in the world's balloon network.

At the time when radio sounding of the upper air first began to be introduced to routine meteorological operations during the 1940s, the importance of obtaining measurements above the oceans was fully realized. As it happened, these were also the early days of transatlantic civil aviation. With flying still a rather chancy affair by today's standards, and aircraft having comparatively limited ranges of operation, special air-sea rescue arrangements were needed. Thus there came into being a network of ocean weather stations over the North Atlantic. Specially equipped ships, provided by several nations in North America and Europe, did their best to keep close to their appointed station come wind or wave, and carried out synoptically coordinated radiosonde launches very much as at land stations, including the radar tracking of the balloons. The ships remained on station for several weeks at a time before being relieved by other vessels. If an aviation emergency should arise, the ships stood ready to take on their

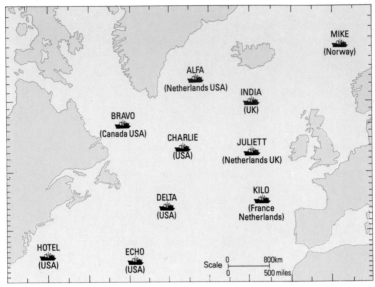

The North Atlantic network at its height during the 1950s, included ten ocean weather stations operated by six countries and making routine radiosonde observations.

air-sea rescue role. The North Atlantic network was fairly quickly developed to the point where it included ten ocean weather stations, and this amounted to a very useful provision of upper air information over a major ocean. A few ocean weather stations were established elsewhere, notably in the eastern North Pacific, but no other ocean area had a comparable coverage.

The North Atlantic ocean weather station network at its height represented something approaching the meteorologist's ideal solution to the problem of knowing what is going on at upper levels in the air above the oceans. But even in the North Atlantic it has not proved to be a lasting solution. As aircraft became more reliable and their ranges extended, the justification for air-sea rescue stations diminished. And viewed solely as meteorological platforms the ships have come to look increasingly expensive. In general the European countries, for whom the North Atlantic is usually upstream, have been willing to fund the operation of ocean weather stations as long as ships remained serviceable, but the capital costs of replacing ships have presented severe difficulties. Inevitably, then, the number of fixed ocean weather stations has dwindled.

But the need for upper air information over the oceans is as pressing as ever. Satellites now play a major role (see pp. 107–114). Yet balloon soundings remain the only well-tried way of detecting the detailed structure of wind, temperature, and humidity profiles – satellite-derived information takes the form of weighted averages over rather deep layers of air. Thus an important recent development is the Automated Shipboard Aerological Program (ASAP), which has established methods of launching radiosonde balloons from merchant vessels and research ships during their normal voyages with comparatively little manual effort.

As far as the temperature, pressure, and humidity measurements are concerned, the ASAP system is much like a conventional radiosonde except that data collection by satellite is a standard provision. A greater innovation has been required for wind finding, however, since there is no way that conventional tracking radar could be installed acceptably on merchant ships. Wind speed and direction are still deduced from a sequence of balloon positions at frequent intervals, but the positions must now be determined remotely. This is done with adequate accuracy by using one of the world's navigational systems. The balloon's horizontal position is determined by receiving the pulsed signals transmitted simultaneously from radio beacons situated at widely separated points on the Earth's surface.

The ASAP systems provide a set of mobile ocean weather stations in place of the old fixed network that was limited to the North Atlantic. As well as their direct value for forecasting, the ASAP soundings are essential for the calibration of satellite measurements over the oceans.

Aircraft reports

With meteorologists ever hungry for more data on the weather machine in action, the expansion of civil aviation in recent decades has provided a valuable additional source of upper air information. The aircraft carry instruments which enable wind and temperature to be determined for the air through which they are passing. Of course a single aircraft can provide information at its own flight level only, so an individual aircraft report is of limited value compared to a balloon sounding. On the other hand, the sequence of aircraft reports during a flight can provide a density of measurements in the horizontal that is unmatched by any other meteorological observing system. And the value of single-level aircraft data is increased by the fact that the flight levels of modern airliners (10–13km/30,000–40,000ft), though chosen for aerodynamic reasons, correspond quite closely to those of the very important jet streams in the atmosphere.

The conventional system for obtaining meteorological information from aircraft is to have wind and temperature values noted by aircrew at agreed reporting points along the route, and included in the standard messages radioed to air traffic control centers. Here the weather data can be extracted from among the other information and fed into communications circuits leading to weather forecasting offices.

In some parts of the world the density of air traffic is such that the number of aircraft reports is sufficient to provide a good description of the atmosphere at flight levels. Over the North Atlantic, for example, the reporting practice ensures that measurements are available whenever the aircraft cross certain lines of longitude spaced at 10° intervals and, for an increasing number of flights, at 5° intervals. The timing of these aircraft reports is deter-

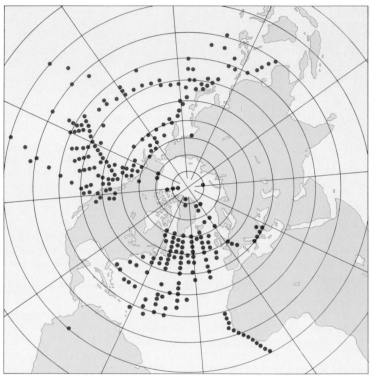

Airplanes provide valuable data at flight levels, mainly over the oceans. The map shows the distribution of reports received for a typical six-hour period.

mined by the flight schedules and the ground speed of the aircraft. These are not synoptic data: there is no way of arranging that all observations refer to standard times – except by ignoring much of the available data. As will be seen, much of the satellite information too is intrinsically asynoptic, i.e. unsynchronized. For forecasting purposes it is necessary to extract as much value as possible from every piece of information, whether synoptic or not.

A fully automated system for obtaining aircraft data for forecasting has been successfully proven in trials and will gradually become more widely used. The information reaches forecasting centers via an Aircraft to Satellite Data Relay (ASDAR). The system, installed on wide-bodied aircraft, provides measurements every $7\frac{1}{2}$ minutes by extracting data from the aircraft's navigational system and storing them. The data stored are one-second samples of wind speed and direction, air temperature, an index of turbulence, and the aircraft's height and position. Bulletins containing eight such sets of measurements are transmitted every hour in level flight and can be available quickly at forecasting centers via the satellite data relay.

During the ascent and descent phases of a flight, the ASDAR mode of reporting is changed automatically and up to 30 observations are made between the ground and cruise level. These data form in effect a slanting equivalent of a radiosonde ascent, and potentially provide a valuable addition to the upper air profiles obtained by using balloons. Near a busy airport one can envisage a more or less continuous availability of up-to-the-minute vertical profiles of wind and temperature – certainly an exciting prospect for forecasters.

The role of satellites

The first artificial Earth satellite was launched in 1957. Within another three years the era of *meteorological* satellites began, opening up new possibilities for monitoring the weather from space. The satellites achieve a regular coverage of the whole globe that would be impossible in practice with surface platforms, balloons, and aircraft.

The most obvious impact of satellites on meteorology is the availability of pictures (often called images) which reveal where cloud is present and what it is like. Since virtually all the important weather systems of the world have distinctive cloud patterns, the satellite images give forecasters the chance to *see* what is going on, even in remote areas, at frequent intervals.

Even more important for modern computer-based forecasting has been the development of techniques to derive temperature and wind data from satellite measurements. In the vast ocean regions of the southern hemisphere these data have made forecasting for a few days ahead a realistic proposition for the first time.

Although satellites give a far better horizontal coverage than other observing systems, satellite data on winds and temperatures are generally not so accurate as those from balloons and aircraft. And certainly the vertical detail that can be detected is much less than from a radiosonde ascent. So forecasters still need all these complementary types of observation.

Satellites make another important contribution to the monitoring of the weather by providing data collection, communication and location facilities for other observing systems. Buoys, automatic weather stations, new systems on ships, aircraft, balloons: satellites are often essential to the effectiveness of all these types of observation.

Before taking a closer look at the information obtained from satellites, it is important to distinguish two types, each with its characteristic orbit around the Earth.

Geostationary satellites always remain above the same spot on the Earth's surface. This is possible only if the satellite is over a point on the equator and if its distance above the surface is about 36,000km (22,400 miles). From this vantage point the satellite can cover a disk-shaped area extending roughly 60° north, south, east, and west of the sub-satellite point. A great advantage of geostationary satellites is that this field of view can be monitored continuously. It turns out that, with the necessary overlaps, five geostationary satellites spaced around the equator can monitor the whole belt from 50°N to 50°S.

Polar-orbiting satellites fly much nearer the Earth's surface and so

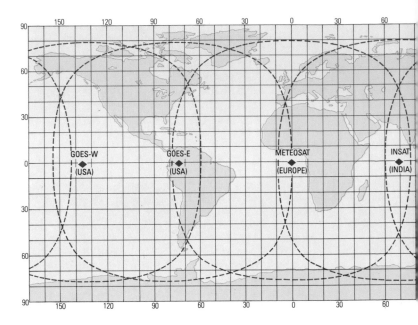

can provide more detailed imagery. Their orbits take them over both the Arctic and the Antarctic regions, and, with the Earth rotating below, the satellite views a different strip of the surface each time around. Typically, a polar-orbiting satellite is 850km (530 miles) above the surface and takes about 100 minutes to go once round the Earth, which meanwhile rotates through some 25°. Often the instruments scan from side to side, so that each part of the Earth's surface is observed at least once in each twelve-hour period by a single satellite (more frequently at high latitudes where the orbital strips overlap). Two such satellites are needed for operational forecasting, giving four observations each day for every area.

Along with good coverage, another strength of satellite observations is that the whole globe can be monitored with a single set of instruments. This avoids problems that arise when, for example, different instruments are used in different countries for balloon measurements. The other side of this particular coin is, of course, that when a satellite instrument fails there is a worldwide loss of information.

Satellite imagery

So how do satellites, hundreds or even thousands of kilometers up in space, provide information about what is going on below? They do so by carrying instruments call *radiometers* that measure radiation of various kinds reaching the satellite from the Earth, the air or the clouds.

One kind of radiation reaching the satellite from below is light from the Sun that has been reflected back by land, sea or cloud surfaces. Each type of surface reflects different proportions of the

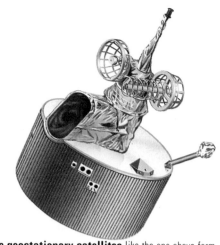

Five geostationary satellites like the one above form a ring above the equator and collect data from the large areas shown (left). Imagery is provided for slightly smaller areas of coverage.

light falling upon it. As the radiometer on the satellite scans the field of view below, it records the intensity of the radiation received from each small area. This information is transmitted back to a ground station, and can be processed to form the familiar "satellite picture" or, as meteorologists prefer to say, "visible image."

Visible images portray cloud, land and sea areas much as the human eye would see them. But of course they can see nothing at night or, in winter, near the poles. But another kind of radiation reaching the satellite from below has the advantage of providing images day and night, all year round. This radiation is heat rather than light, and, strange as it may seem, this is emitted by all surfaces whether hot or cold. The radiation of this kind from the Earth and from clouds is known as infrared.

Infrared images depict the temperature of the surface viewed. Measurements from an infrared radiometer are processed at the ground station, usually in such a way that warmer surfaces appear relatively darker and colder surfaces relatively lighter in the resulting black-and-white image. So land usually appears darker than sea in summer but lighter in winter. Cloud shows up very clearly, provided that the cloud top temperature is sufficiently lower than the temperatures of the Earth's surface in the surrounding cloud-free areas. The infrared measurements can, alternatively, be color-coded to highlight temperature contrasts.

At night the infrared image allows the forecaster to see many very important cloud patterns. In the daytime the availability of simultaneous visible and infrared images is even more effective. For example, clouds that may look very similar in the visible will often show up with marked differences in the infrared because one

Simultaneous visible (above left) and infrared (above right) images from a polar-orbiting satellite. In the visible, all areas of cloud, fog, sea ice and snow cover show up because of their greater reflectivity. In the infrared the high cloud tops are picked up because of their lower temperatures.

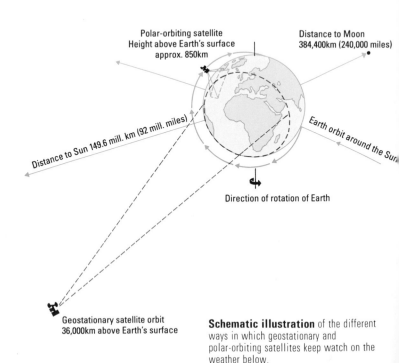

Polar-orbiting satellite
Height above Earth's surface approx. 850km

Distance to Moon
384,400km (240,000 miles)

Distance to Sun 149.6 mill. km (92 mill. miles)

Earth orbit around the Sun

Direction of rotation of Earth

Geostationary satellite orbit
36,000km above Earth's surface

Schematic illustration of the different ways in which geostationary and polar-orbiting satellites keep watch on the weather below.

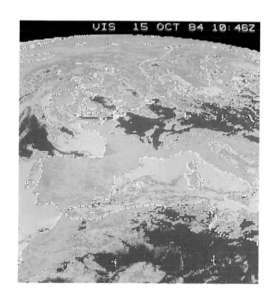

Infrared or visible radiation measurements received from a satellite may be color-coded to produce a false-color image (left). The color-coding can be chosen so that features of special interest are highlighted.

is much deeper, with a much colder top, than the other. On the other hand, areas of very low cloud or fog will escape notice in the infrared (because their temperatures are so similar to those of surrounding clear areas), yet will show up clearly in the visible.

Infrared measurements can do more than provide images of cloud distributions. From the intensity of the radiation received at the satellite, the numerical value of the temperature of the radiating surface can be calculated. Information on the slowly varying temperature of the sea surface is obtained in this way and, as will be seen, is important for forecasting, especially for longer-range forecasts.

Visible and infrared imagery is available from both geostationary and polar-orbiting satellites. Both have advantages. The polar-orbiters are much nearer and provide more detailed images, made up of individual elements one to two kilometers (around a mile) in size. On the other hand, geostationary satellites provide images of the same region at frequent intervals, usually every half hour.

Using a pair of geostationary images half an hour apart, the movement of an identifiable cloud element can be measured. If that cloud is being carried along by the air, the movement shows the wind speed and direction averaged over half an hour in the area through which the cloud has moved. Hundreds of these "cloud motion winds" are calculated every day by automatic processing of geostationary images, sometimes with some human monitoring for quality control. It is also necessary to know the level at which the wind has been calculated, and this can be estimated from the cloud top temperature (obtained from infrared measurements).

Cloud motion winds are greatly valued because they provide numerical data in regions over the tropical oceans where little other information is available. However, forecasters have to treat them carefully because they are subject to errors. One common source of error is the estimation of the level. Another is misinterpretation of clouds which are *not* moving with the air; for example, cumulus clouds may be anchored to warm spots on the surface, and stationary wave clouds may be present downstream of mountains.

Other satellite measurements

Perhaps the most ambitious of all the measurements made using weather satellites are those designed to provide numerical values of *air* temperature at various levels. The air is a mixture of gases, and each gas emits infrared radiation of particular kinds. The satellite radiometers used for temperature soundings are usually tuned to measure the radiation from the small but well-known quantity of carbon dioxide that the air contains. By measuring the intensity of radiation at a number of different wavelengths, and by making complicated calculations based on radiation physics, it is possible to derive several values of the air temperature, each averaged over a different layer. And although lacking the vertical detail of balloon soundings, the satellite soundings provide upper-air temperature information where otherwise none would exist, including all the ocean areas.

In cloudless air the satellite sounders on polar-orbiters can provide data at intervals of about 80km (50 miles) both along the track and across swathes extending more than 1,000km (620 miles) either side of the track. Clouds are a big problem, though, because all infrared radiation from air below the cloud top is blocked off. Every effort is made to use gaps in the cloud, but inevitably there is a poor coverage of infrared soundings in cloudy regions, which is serious because these are often the most important regions for forecasting purposes. To help with this problem satellites are also fitted with radiometers which measure the microwave radiation emitted by the air. Fortunately, microwave radiation can pass through most clouds unscathed.

The humidity content of the air can be estimated from satellites using radiometers tuned to wavelengths characteristic of the infrared radiation emitted by water vapor. Given that the temperature is known, the intensity of the radiation depends on the quantity of water vapor present.

Vertical sounders have been installed on some geostationary satellites, but their main role has been on polar-orbiters, for which there are two methods of data collection. The data are recorded on board the spacecraft, and then the measurements from one or more orbits are transmitted when passing over designated ground stations. However, the data are also transmitted continuously as they are measured, and a ground station can immediately receive these "direct read-out" measurements for its local area whenever it has line of sight with the satellite. Data received in this way provide the input to Local Area Sounding Systems (LASS), which produce upper-air temperature information on a regional basis.

A more recent development in satellite instrument technology has been the use of *active* remote sensing. Rather than passively

Cloud motion tracked by geostationary satellites can reveal wind, but not when clouds are anchored to surface heating, as (left) along the Atlantic coast of Florida in August.

Temperature soundings (below) from polar-orbiting satellites provide southern hemisphere coverage that would be impossible by any other method. The map below shows the soundings received for a six-hour period from a single satellite.

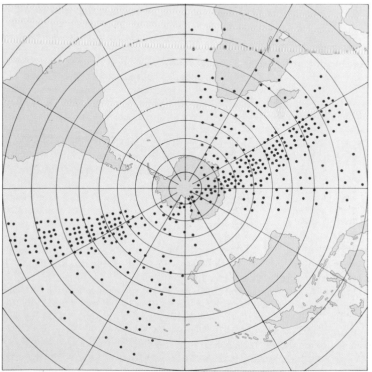

measuring radiation reaching them from below, these instruments transmit carefully controlled signals of their own, and measure what is reflected back. Such instruments have important applications to monitoring the sea surface. For example, a microwave instrument called a scatterometer can detect tiny ripples on the water surface, superimposed on the waves and swell, and the measurements can be used to estimate the surface wind speed and direction. The height of the waves themselves can be deduced, using an instrument known as a radar altimeter.

There seems almost no limit to the weather information that might be obtained from satellites as increasingly ingenious methods are devised. Doubts remain, though, about the accuracy that can eventually be achieved. Some of the ideas that are coming up would be very costly to implement, but their advocates are convinced that the benefits from improved forecasts would far outweigh the costs.

Radar and rainfall

The surface stations, ships, buoys, balloons, aircraft, and satellites mentioned in previous pages provide worldwide information on wind, pressure, temperature, humidity, and cloud. But what about measurements of rainfall? These are certainly important, particularly when it comes to short-period forecasting (i.e. a few hours ahead), for which a close watch must be kept on the movement and development of rain patterns. Simple visual observations of the occurrence and intensity of rain (or other kinds of precipitation) are valuable in themselves, and quantitative information can be obtained with the familiar rain gauges. Telemetering rain gauges, capable of transmitting data, are used when the measurements are needed quickly at a central forecasting office.

Because rainfall varies so much from place to place, especially near coasts or hills, it is often impossible to obtain a full picture of what is happening – even in regions where a dense network of rain gauges is available. Weather radar is a much more powerful technique for rainfall measurements. With the radar beam scanning horizontally, any rain, snow or hail that is within range will reflect the signal, offering a detailed plan view of the distribution of precipitation.

The intensity of the radar echo received back from raindrops can be related to the rate of rainfall with the help of a few telemetering rain gauges for calibration. However this quantitative information is increasingly unreliable at ranges beyond about 75km (48 miles). Radar displays are typically limited to a coverage of 150km (96 miles) compared with maximum operating ranges of around 210km (130 miles).

Since a radius of 150km (96 miles) is too small for all but the most local weather forecasting, weather radars should ideally be operated in networks. The measurements from each radar in the network can be processed by a central computer. After calibration and quality control have been carried out, the measurements are color-coded according to rainfall intensity, and a composite image for the whole area is prepared for the use of forecasters.

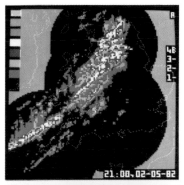

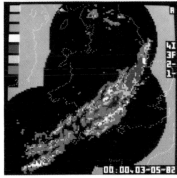

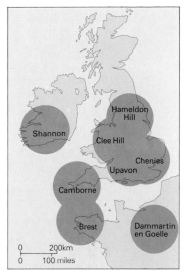

A network of weather radars (above) can provide unique information on detailed rainfall patterns. The examples on the left show how the composite images from the network can be used to monitor the advance of cold front rainfall at frequent intervals.

Since the radars provide continuous surveillance, such images can be provided at frequent intervals – every 15 minutes is typical.

The effective region for which information on rainfall is available can be further extended by combining measurements from radar with those from geostationary satellites. Although the satellite imagery depicts cloud rather than rain, there is often a close correspondence between deep clouds with cold tops, easily identified in infrared satellite imagery, and the areas of heavy rainfall revealed by a radar network. By studying the distribution of cloud and rain in the area with satellite *and* radar coverage, useful inferences may be drawn about rainfall in the surrounding areas which have satellite coverage only.

Experimental observing periods

Monitoring the global weather and providing the data necessary for forecasting are difficult tasks. Every available piece of information from whatever source must be used to the full. Even then there are gaps in the coverage, and this often means that the total information routinely available is insufficient for some aspect of understanding or forecasting. So there is a need for experimental observing periods when additional measurements, which could not be afforded as routine, are made in an attempt to clarify how some aspect of the weather works. The temporarily improved coverage of measurements provides a basis for research which may

Specially modified research aircraft with instrument-carrying appendages have an almost comical appearance. They play roles in experimental observing periods but are too costly to help with routine measurements.

take several years to follow through. If this results in improved understanding, then weather forecasts can be expected to improve too. An experiment may also show that additional deployments of observing systems operationally would be cost-effective.

The biggest experiment ever organized took place from December 1978 to November 1979. Known as the Global Weather Experiment (GWE), its aim was to improve the accuracy of weather forecasts for a few days ahead in all parts of the world. Several of the observing techniques mentioned in earlier sections were first implemented under the stimulus of the GWE. It was then that the set of five geostationary satellites around the equator was established for the first time, and wide-bodied jet aircraft were fitted with equipment for the Aircraft to Satellite Data Relay (ASDAR).

For the GWE special attention was given to observations in the tropics and the southern hemisphere. Weather ships making balloon soundings were stationed in the tropics, and at certain times of the year took up positions of special relevance to the Asian and West African monsoons. Research aircraft flew along pre-planned routes and dropped instrument packages which descended by parachute (*dropsondes* – similar in operation to the radiosondes carried by balloons). Constant level balloons were launched which drifted with the wind at around the 14km (46,000ft) level. Each carried a radio transmitter so that it could be located by polar-orbiting satellites, and wind speed and direction were determined from each pair of successive positions. A greatly enhanced coverage of drifting buoys was achieved in the southern oceans, and resulted in new discoveries of basic facts

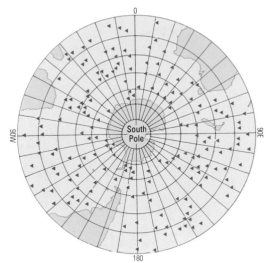

Fully instrumented buoys (above) are deployed in coastal waters, but large numbers of simpler buoys were used in the Global Weather Experiment, and provided unprecedented coverage (left) of surface pressure data over the southern oceans.

about the intensity of low pressure systems in the southern hemisphere.

The GWE had a major impact on operational forecasting not only through the development of new methods of observing but also through improvements in the techniques of numerical prediction which will be described in the next sections. The experience gained during the GWE was important in planning the next stage of international collaboration in research on the global scale – the project known as the World Climate Research Program. Here the emphasis has shifted to monthly and longer time scales, where observations of motion, temperature and salinity in the oceans are needed in conjunction with atmospheric measurements.

Another example of an experimental observing period, this time with a regional emphasis, is the Genesis of Atlantic Lows Experiment (GALE), which involves intensive observing near the east coast of the USA during the period mid-January to mid-March. During a typical winter, low-pressure systems may develop frequently in the region, at the confluence of warm air from the tropical Atlantic and cold air from the Canadian Arctic. Some of these storms develop rapidly and, as they move north and east, may bring sudden blizzards to the densely populated zone from Washington to Boston. These dangerous blizzards have proved difficult to forecast over the years, so GALE was organized to obtain sufficient observations for a detailed investigation of the mechanisms involved. During the GALE observing period in 1986, some 50 stations made balloon soundings of the upper air at *three-hourly* intervals, and eight research aircraft were used to measure the fine structure of the wind field. With two research vessels and eight special buoys also contributing data, GALE has set new records in its sheer concentration of resources on a particular weather problem.

Putting the computers to work

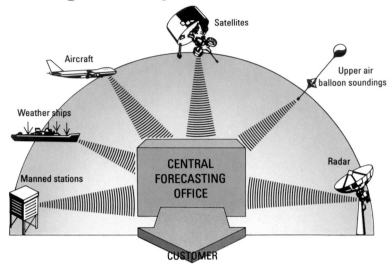

Surface stations, ships, airplanes, satellites, radiosondes and radars all contribute to the composite global observing system that feeds the weather computers.

We have seen how thousands of observations from many different sources take place each day all over the world. But what then? How can the weather be forecast from these observations?

First, the observations must be gathered together. Surface observations and balloon soundings are sent to local collecting centers and then on to national centers; measurements from aircraft, ships, buoys, automatic weather stations, and balloons are collected via satellites; satellite measurements themselves are received at processing centers, where imagery is prepared and temperatures or winds are calculated. Then, after these initial stages of collection, a free international exchange of information takes place, using a network of special links for meteorological purposes only, the Global Telecommunications System (GTS).

When sufficient observations have arrived at major forecasting centers, they are fed into powerful computers programmed to carry out calculations known as "numerical weather prediction." This is basic to modern forecasting. The next section explains how the computers are put to work.

Forecasting by numbers

The atmosphere is measured at locations distributed irregularly around the surface of the Earth, and, though some of the measurements are synoptic or "seen together," many take place at differing times dictated by aircraft flights and satellite orbits. By contrast, computer weather forecasts must start from synoptic values of wind, pressure, temperature, and humidity as a regular array of horizontal locations, known as *gridpoints*, and at a fixed set of *levels* in the vertical. Just how the synoptic gridpoint values are calculated from the measurements that have been gathered together – a process known as analysis – is a crucial part of the story that will be outlined later.

Once the analysis has been completed, it becomes possible to apply mathematical equations that represent all the important physical processes, and so to calculate the change that will occur in each gridpoint value during a short interval of time called the *time-step*. The calculation at each gridpoint will involve additions, subtractions, and multiplications using values from the surrounding gridpoints. Then, when new values of the variables (wind, pressure, temperature, and humidity) have been calculated for all the gridpoints at all the levels, the whole process can be repeated to move another step forward in time. In this way a forecast for a few hours, a day, several days, or even longer ahead, can eventually be produced.

These calculations of wind, pressure, temperature, and humidity have important by-products. For example, if the humidity calculated at a gridpoint, at a particular level, rises above a certain value (dependent on pressure and temperature), a further stage in the calculation is invoked whereby some of the water vapor is converted into liquid water, or to ice if the gridpoint temperature at that level is below freezing. This results in a forecast of precipitation be it rain or snow at the gridpoint.

Calculations of this kind, based on gridpoint values, are approximations of the truth: very good approximations, but approximations nonetheless. The closer together the gridpoints, the better are the approximations, but also the greater is the number of additions, subtractions, and multiplications that have to be calculated. And if the gridpoints are closer, the mathematics requires that the time-steps must be shorter too.

For accurate computer forecasts, a gridpoint spacing of at most 100km (62 miles) is desirable, with 20 levels in the vertical. For a global forecast that means about 40,000 gridpoints, and with four variables at each level it follows that some $3\frac{1}{2}$ million numbers are needed to specify the state of the atmosphere. The corresponding time-step is about 10 minutes, so a global forecast for one day requires about 500 million calculations of gridpoint values, with each calculation involving a whole set of basic arithmetic operations. No wonder that powerful computers are needed to keep the forecast ahead of the weather.

The system of equations, approximations, and calculations is called a numerical *model* of the atmosphere. While the primary role of numerical models is to enable computer forecasts to be made, they are also valuable tools for research into weather and climate.

Numerical models of the atmosphere for global forecasting require a large number of calculations to be made in a short time. Often the computing power available is not sufficient, and one of two things must happen. Either the model remains global but the spacing of the gridpoints is increased to reduce the computational workload, or the fine mesh of gridpoints is retained but the model is restricted to a limited area of the Earth's surface.

An advantage of limited-area, fine-mesh models (LFMs) is that observations from the limited area reach the forecasting center earlier than those from distant parts of the world. The LFM can

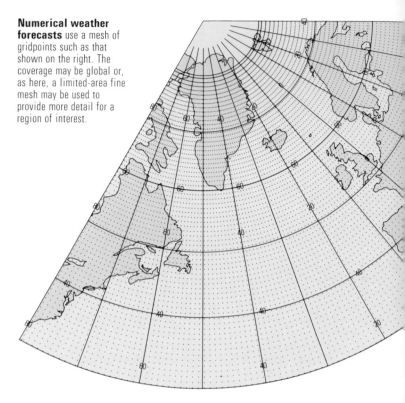

Numerical weather forecasts use a mesh of gridpoints such as that shown on the right. The coverage may be global or, as here, a limited-area fine mesh may be used to provide more detail for a region of interest.

then be used to produce an early forecast for the region, with valuable hours saved compared to later global forecasts. At several centers, both global models and LFMs are in routine use. Whereas a global model can provide useful forecasts for about a week ahead, LFM forecasts are usually limited to 36 or 48 hours ahead. At the gridpoints on the boundary of the limited area, calculations using values at neighboring gridpoints cannot be performed. Instead, information from a global model is used to update these boundary gridpoints. Sometimes, for a small area of special interest within a LFM, another model with even closer spacing of its gridpoints, say 10km (6 miles), is used to provide more detailed forecasts of cloud and precipitation, probably just for 12 or 18 hours ahead.

The analysis process

For all these types of numerical forecasting the analysis, that is the calculation of the starting values of the variables at the gridpoints, is an essential first step. At each gridpoint and at each level the starting values of wind, temperature and so on are calculated, using whatever measurements are available within a chosen radius of influence and within a chosen interval of time. The gridpoint value is calculated as a weighted average of the measurements, with each measurement given more or less weight according to its

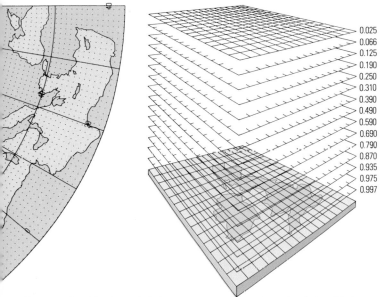

| 0.025 |
| 0.066 |
| 0.125 |
| 0.190 |
| 0.250 |
| 0.310 |
| 0.390 |
| 0.490 |
| 0.590 |
| 0.690 |
| 0.790 |
| 0.870 |
| 0.935 |
| 0.975 |
| 0.997 |

Above each gridpoint, calculations are made at 15 different levels in this particular model. The levels are labeled with values of pressure expressed as a fraction of the pressure at the surface below. The lowest level (0.997) is about 25m (82ft) above the surface and the highest level (0.025) about 25km (82,000ft) above the surface.

distance from the gridpoint, its separation in time from the starting time of the forecast, and its expected accuracy. The use of the measurements is made much more effective by combining them with a background of forecast gridpoint values from an earlier starting time. This can be done by applying the weighting factors, not to the measurements themselves, but to the differences between the measured and the background forecast values.

Each type of measurement has its expected range of errors, and these are taken into account in determining the weighting factors for the analysis, as indeed are the expected errors of the background forecast. But some of the observations arriving at a forecast center may be seriously in error, due perhaps to the malfunction of instruments, to shortcomings in data retrieval methods or to transmission errors. Every effort must be made to recognize and reject these rogue measurements so that they do not introduce major errors in analyses and hence in forecasts. Human forecasters can play a useful role in these quality control decisions, as will be seen in a later section.

There are two important attributes that a good analysis of gridpoint values should have, if accurate forecasts are to be obtained. Firstly, the gridpoint values should be faithful to all the measured data except those identified as rogues. Secondly, those of the starting values that are not directly controlled by the available measurements must be realistic. For example, the proper balance between gridpoint values of wind and neighboring gridpoint values of pressure must exist even in regions where pressure measurements are available but wind measurements are not.

It has to be accepted, of course, that there will always be errors in measurements and consequently in the analysed values. Unfortunately the physics of atmospheric motion is such that initial errors, however small, will grow during the forecast. So even though in principle the forecast model could be stepped forward indefinitely, in practice there is a limit to "predictability." In other words, over a certain period the errors in the forecast will have grown to the point where the information is no longer useful. The period of useful predictability for global forecast models with 100km (62 miles) gridpoint spacing and 20 levels in the vertical is about one week, with variations on different occasions from as little as four days to as much as ten days. Future research will probably make it possible for these periods to be extended – perhaps doubled – but the limit will always remain.

Supercomputers

The first person to suggest that forecasting by numbers could work was Lewis Richardson, whose book on the subject was published in 1922. That was before the days of electronic computers, and Richardson's vision was of an array of human calculators, each looking after his or her own gridpoint. Richardson himself carried out a trial forecast at a single gridpoint, but the result was hopelessly wrong, mainly because of a lack of balance between wind and pressure in the starting conditions. Not until the early 1960s was his suggestion finally vindicated in its original form, though a less ambitious type of numerical model has been used with limited success from about 1950 onwards.

The history of numerical weather prediction is in many respects the history of powerful computers. Right from the beginning, meteorologists have been ready and waiting to exploit improvements in computer performance as soon as they are developed. A prime factor has been the need to reduce the spacing between the gridpoints to get more accurate forecasts. If the grid spacing is halved, the number of gridpoints in a given area goes up by a factor of four and, because the time-step must also be halved, the computing workload increases eight times. Other factors have been the increasing number of levels and the inclusion of more and more complex calculations of the various physical processes in the atmosphere.

Some features of numerical weather prediction models make them particularly suitable for certain advanced types of computer. For example, in a global model there may be 40,000 gridpoints where essentially similar calculations have to be made. This is ideal for "vector processors" – computers that can carry out the same arithmetic operation for whole strings of numbers very much faster, pro rata, than they can for a single set of numbers. Or again, numerical forecasts are well suited to "parallel processors" – computers in which the calculations in different sections of the grid can be carried out simultaneously.

As well as needing to be heavyweight in terms of calculating power, the computers for weather forecasting must have large volumes of high-speed memory to hold the vast arrays of winds, temperatures and so on. All in all, these computers are a far cry from the hardware that is familiar in many offices and homes, and they are given a special name – supercomputers.

Supercomputers make it possible for numerical forecasts to be calculated amazingly quickly. The choice of mathematical methods for stepping the forecast forward also affects the computing time, but an efficient global model running on a super-computer can produce a 24-hour forecast of 20 levels with 100km (62 miles) grid spacing in about 10 minutes. This time is short compared to the several hours needed to collect the observations, and indeed the analysis and output stages of the data processing often take longer than the numerical forecast itself.

It seems likely that supercomputer technology will continue to advance. Meanwhile, research on forecasting continues to show the benefits of finer-mesh global models and very high resolution limited-area models. In addition, more expensive computers are needed to process satellite measurements at their full instrumental resolution, and to assimilate the information into numerical models. So the more advanced supercomputers can be put to work just as soon as they are available.

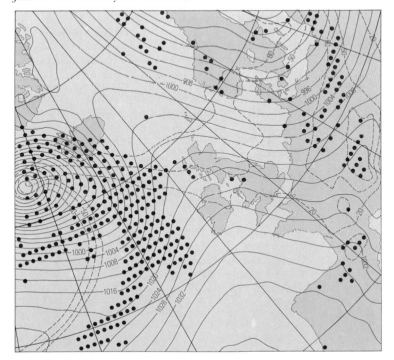

Supercomputer calculations have produced this forecast of sea level pressure (continuous lines) and precipitation (dots at gridpoints). The three dashed lines show forecast temperatures which mark the transition zone from rain to snow.

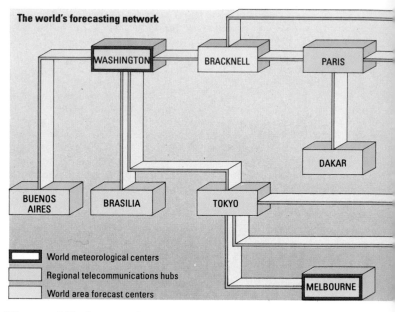

The world's forecasting network

WASHINGTON · BRACKNELL · PARIS · DAKAR · BUENOS AIRES · BRASILIA · TOKYO · MELBOURNE

World meteorological centers
Regional telecommunications hubs
World area forecast centers

The world's forecasting network

The supercomputers and the numerical forecasting expertise necessary for global weather prediction are available at a few centers only. Some customers can be served directly, using results from the global models, but frequently the customer deals instead with regional or national centers. These centers may have a capability for regional numerical prediction, for which information from global models is needed to update the boundary gridpoints. Many regional and local forecasting services are based on forecasters' ability to interpret and apply the numerical forecasts prepared elsewhere.

In all this, the rapid distribution of numerical forecasts to other forecasting centers and to end-users is crucial. Once again, as with the collection of observations, the importance of meteorology's Global Telecommunications System is clear. Every national forecasting center is linked to a Regional Telecommunications Hub (RTH) which is networked with the other RTHs and so can deliver information from any part of the system.

The arrangements set up by the International Civil Aviation Organization provide an instructive example. Two global modelling centers have been chosen, and each acts as a World Area Forecast Center (WAFC). The reason for having two centers is to ensure continuity of service should one be temporarily out of action. The WAFCs compute global forecasts of winds and temperatures at flight levels and of other information relevant to aviation. These forecasts, usually to 36 hours ahead, are distributed as sets of gridpoint values to designated Regional Area Forecast Centers. Each RAFC has three main functions. Firstly, the gridpoint forecasts are passed on to customers for flight planning

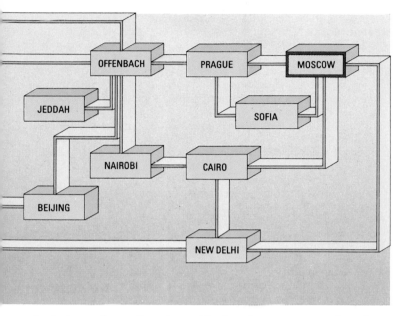

calculations. Secondly, the gridpoint forecasts are plotted on charts and provided as flight documentation to aircrew. Thirdly, forecasters interpret the numerical forecasts and other information to produce "significant weather charts" showing, principally, jet streams and likely areas of turbulence and icing.

A similar system, with world, regional, and national centers, is coordinated by the World Meteorological Organization for more general forecasting purposes. Forecasts of wind, pressure, temperature, humidity, cloud, and precipitation are distributed.

In some parts of the world, specialized centers have been established on a national basis where there is a dangerous threat from a particular type of weather system – severe local storms (including tornadoes) and hurricanes (elsewhere called typhoons or tropical cyclones), are the best known examples.

Specialized centers organized on an international basis have additional advantages in principle from the pooling of financial resources and scientific expertise. Suitable specialities are medium-range forecasts (4–10 days ahead), long-range forecasts (beyond 10 days), monsoon forecasts and drought warnings; but of course these are tasks which present formidable scientific challenges.

Practical considerations have led to a close association between numerical weather prediction and certain types of oceanographic forecasting. Surface wind forecasts are the principal input to state-of-sea models (for predicting waves and swell), and storm surge models (for predicting the departures of tidal levels from their astronomically determined values). Models of the upper layers of the ocean, useful for predicting sea surface temperatures in particular, can also be coupled to numerical models of the atmosphere.

Forecasters and computers

For some applications the forecasts from computer models of the atmosphere can be used directly. A notable example is flight planning for aviation (see pp. 137–138). In other cases, the numerical forecasts become the input for other models, as when forecast surface winds are used to drive state-of-sea models. But for many customers, a forecasting service is provided by the collaboration between forecasters and computers that is known in the trade as "the man–machine mix."

Ideally, forecasters and computers get together in such a way that their differing skills are used to best advantage. Computers are undoubtedly better than forecasters at calculating the changes in airflow that occur from day to day in the atmosphere. On the other hand, forecasters are usually better than computers at deciding if a forecast is going wrong, at recognizing the implications of evidence from a variety of sources, and at making detailed interpretations of the likely weather conditions in particular localities. So forecasters and computers together are more successful than either could be separately.

The weather today
The first area of cooperation between forecasters and computers comes in the analysis of observations to obtain the initial values of pressure, temperature, wind and so on for a numerical forecast. Some of the observations may be seriously in error, and decisions must be made about which to use and which to ignore. Computer calculations can be used to detect some definite rogues, but for borderline cases it is best to bring in human skill. With the help of visual display units (VDUs), forecasters can consider the evidence from other observations, from background forecasts and from satellite imagery, and so make an informed decision as to whether a particular observation is to be believed or not.

Satellite imagery comes into play again, particularly in areas where few measurements are available, if forecasters can recognize from the cloud patterns the presence of a weather system such as a depression or a front which is missing from the computer analysis. Again using VDUs, the forecasters can "intervene" to modify the gridpoint values held in the computer, and to make them more consistent with the cloud pictures.

Some customers want to know what the weather is *now* in a particular locality. This is not as simple a task as it sounds because there are many gaps in the coverage of surface observing stations, and also because the information is always needed very quickly. The techniques used for this work, often called *nowcasting*, are heavily dependent on radar and geostationary satellites. Radar measurements and satellite images are used, for example, to estimate the rate of rainfall now (or just a short time ago) in a river catchment area, so that warnings can be issued to the public if there is a risk of flooding, and to river engineers, who can then assess how to adjust the flow of water if necessary.

Nowcasting techniques may include provision for very short forecasts, just one or two hours ahead, by extrapolating the movement of cloud or rainfall patterns as revealed by sequences of

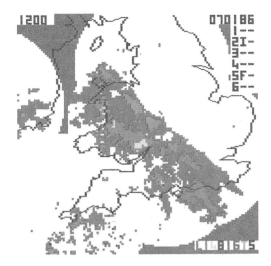

Color-coded radar images (left) were carefully watched by forecasters as this band of rainfall advanced. On this particular January day there was a threat that the rain would turn to snow.

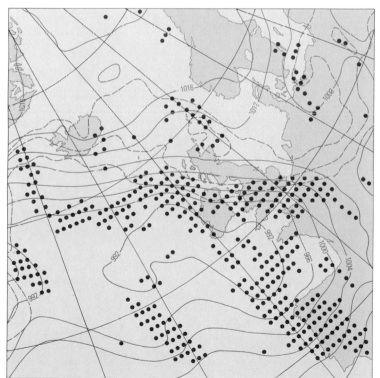

Chart showing the 24-hour forecast available the previous day from the fine-mesh numerical model, valid at the time of the radar image at the top of the page. The forecast rainfall (dots) agreed very well with the radar. The dashed lines warned forecasters that the transition to snow would not be far.

satellite or radar images. Forecasts for a little longer, say 2–12 hours ahead, sometimes pose a special problem. During such a period there may not be any major changes in the airflow, but very often there will be important changes in the detailed meteorology that are difficult to predict accurately. One example is the case of air at low levels moving overland from a cool sea, as frequently happens on the west coast of North America and, in Europe, around the coasts of the North Sea. The moisture picked up from the sea surface is often confined to a shallow cloudy layer. The forecaster has to decide whether this cloud will "burn off" overland during the daytime as a result of heating from the Sun. Some days it does and some days it does not. The difference may be only a few tenths of a degree in the temperature of the air near the top of the cloudy layer. But the effect that can have on the weather is enormous. In one case the day stays overcast and cool; in the other, the sky clears and the surface temperature rises sharply.

The main hope for improving the accuracy of these finely balanced aspects of short-range forecasts lies with very high resolution numerical models, say with 10km (6 miles) grid spacing, and with many levels close to the Earth's surface. These are known as mesoscale models, because they are designed to deal with the middle or "meso" scales of motion that lie between large-scale depressions and anticyclones nad small-scale cumulus clouds. Clearly, these models have to include detailed calculations of the various processes acting to change the temperature and humidity and cloud content of the air. The nowcasts obtained from radar measurements and satellite imagery, with the assistance of quality control and intervention by forecasters, provide the essential starting conditions for mesoscale numerical forecasts.

The weather tomorrow

In some parts of the world, including most tropical regions, the weather tomorrow is likely to be the same in general character as the weather today for long seasons of the year. There are of course variations in detail – for example, in the timing and location of groups of thunderstorms – and these are very difficult to forecast.

There may also be the risk of dangerous developments such as hurricanes in certain seasons, or the problems of forecasting a change of season, as with the onset of monsoons. But for many days of the year in many places, tropical forecasters will require advice from mesoscale numerical models, supported by observations from radars and geostationary satellites, before more accurate forecasts of the daily detail can be expected.

At higher latitudes in both hemispheres the weather tomorrow is quite likely to be radically different from the weather today because of the movement of the weather systems. If the center of a depression passes by, the wind will change from southerly to northerly (taking the case of the northern hemisphere) with a corresponding fall in temperature. If a cold front goes through, a warm, moist, southwesterly flow ahead of the front will give way to a cooler, clearer and possibly showery airstream afterwards,

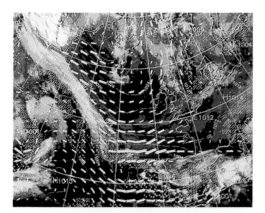

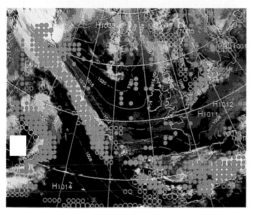

Modern meteorological software provides screen displays of computer calculations superimposed on satellite images. First (above), the computer's *low-level winds* are viewed with the cloud pattern (as seen in the infrared).

Next the clouds are compared with the computer's *rainfall* (left). Agreement is good along the front in mid-Atlantic, but a small positional error in the computer's rainfall is shown up further west.

The same satellite image is displayed again, and this time the forecaster has selected the *sea level pressure* pattern for comparison (below). These techniques are used to monitor the computer analysis and also to check the early stages of the numerical forecast.

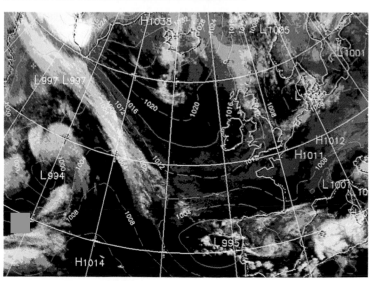

probably with heavy precipitation at the time of the transition.

Numerical forecast models, especially the limited-area fine-mesh models, make a major contribution by predicting the movement and development of depressions and fronts with impressive accuracy. Forecasts for 24 hours ahead, which only 20 years ago were notoriously unreliable, can now be trusted in the great majority of occasions. And the models provide a fairly comprehensive set of information on the broad scale – pressure patterns, precipitation, wind speed and direction and so on.

An important role for forecasters in collaboration with these generally successful computer predictions of the weather tomorrow is to assess the accuracy as soon as possible after the calculations have been completed. One of the most useful techniques is to compare the early stages of a computer forecast with satellite imagery for the same hour. Time is precious for this work – a weather forecast is a highly perishable commodity – and it is a great help if forecast fields can be superimposed upon the imagery on a VDU screen.

Sometimes the comparison of numerical model fields of precipitation or cloud with contemporary satellite imagery may reveal that a major error is developing in the forecast. Perhaps the model is growing a new depression on an already existing front, but the satellite information does not show the characteristic bulge of deep cloud. In extreme cases the forecaster may conclude that the numerical forecast should be disregarded entirely.

Much more frequently, the comparison of model and satellite data shows that, while the developments and movements indicated by the early stages of the forecast are correct, there is an error in timing. An advancing rainband may be found to be a few hours ahead or a few hours behind its forecast position. The advice to be provided to customers, based on the later stages of the numerical forecast, can then be adjusted accordingly.

Adding local detail

Once the general pattern of weather to be forecast for tomorrow has been settled, the stage is set for a further aspect of the cooperation between forecasters and computers. This concerns the interpretation of the general pattern in terms of particular aspects of the weather and with attention to local detail.

Considering precipitation for example, satellite and radar displays reveal the complexity of the detail which the forecaster must interpret, as far as possible, from the numerical guidance. Often a forecaster can do little more than advise that considerable local variations will occur, as with the showery conditions that frequently characterize northwesterly airstreams coming ashore from the Atlantic or Pacific oceans.

With experience, some specific details can be added, as for example when showers form inland on summer afternoons while the upwind coastal strip enjoys unbroken sunshine. Hills, valleys, and lakes impose important local detail on the distribution of showers, and many fronts have patchy precipitation patterns.

In cold weather, an interpretation is required regarding the

Forecasters use their experience to interpret computer outputs and other information in the final stages of advising a special customer or adding local detail to a general forecast.

form – rain or snow – of any precipitation that is forecast. General guidance is provided by the temperatures of the lowest layers of air as calculated by the numerical model. Studies of past data have been used to relate these temperatures to the probability that any precipitation will be in the form of snow. Once again, there is considerable local variation, particularly that arising from differences in altitude, with the lower surface temperatures over higher ground increasing the probability of snow.

The interpretive stage is required for every weather element that is forecast; maximum and minimum air temperatures, ground frost, cloud and fog, gales and so on. The experience built up by practicing forecasters is essential, but automated assistance can also be provided. The mesoscale forecasting models may be regarded as sophisticated interpretive tools, adding greater detail to the regional models within which they are set. Statistical relationships between long series of past observations and the corresponding numerically forecast values can often capture local detail when applied to tomorrow's forecast.

Next week's weather

Numerical forecasting using supercomputers has made a major contribution to the improved accuracy of weather predictions for one or two days ahead. Perhaps even more revolutionary has been its impact on predictions for three, four and five days ahead. Records of the errors in sea-level pressure forecasts for North America and Europe show that in the mid-1980s three-day forecasts were, on average, as good as one-day forecasts had been ten years before, with five-day forecasts as good as three-day forecasts had been. Of course, the accuracy inevitably diminishes as the forecast period is extended. But even at seven days ahead

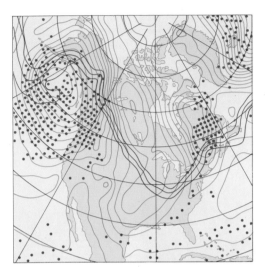

The surface pressure pattern over North America on this January day showed cold, dry air pushing south of the Great Lakes on the eastern flank of a high pressure system. Rain and snow from low pressure systems were affecting the West Coast and eastern Canada.

The five-day computer forecast (right), produced the same day as the map above, correctly indicated that the cold air would move much farther south, with a strong northerly flow developing all along the East Coast. The lines marking the transition from rain to snow have reached Florida in the forecast.

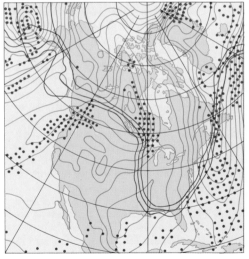

numerical models provide useful guidance to forecasters on many occasions. For these longer periods the numerical models need to be global in coverage because, in the space of a week, the forecast for a particular region may be influenced by weather systems on the other side of the world. Examples are the development of a depression near Japan that can have a downstream effect on European weather a week later, and the onset of the southwest monsoon over India which may be affected by earlier mid-latitude developments taking place in both the northern and southern hemispheres.

Forecasting for a week ahead differs from forecasting for a day ahead in that it is no longer possible to predict details of the weather with confidence. Whereas every effort is made, one day ahead, to forecast the exact position of features such as depressions and fronts, it is not realistic to attempt similar precision in the forecast of next week's weather. In particular, there are likely to be errors in the detailed positions or shapes of any newly formed depressions or fronts. These may come into existence during the course of a forecast, with important consequences for the weather at the end of the period.

So forecasters have to be guarded about details of the weather a week ahead. What they *can* reasonably attempt, using the results from global numerical models, is to predict the general *types* of weather that will occur. For example, the flow in winter may be such as to bring very cold air into a particular region, as when air reaches Florida and other southern states from the Canadian Arctic, or when an easterly type brings air from Russia across western Europe. At the same season in other years, flows from other directions bring totally different conditions to the same regions. This accounts for much of the variation of weather from year to year in a given month – variation that often perplexes the general public.

As well as the differing origins of the air reaching a particular region, another important distinction is that between changeable and settled weather types. A typical changeable type occurs when several depressions form in succession and move into North America from the Pacific or into Europe from the Atlantic. Periods of rain and showers alternate with dry and sunny days, with the wind also likely to change in direction and strength from one day to the next.

Settled weather conditions occur when a large-scale flow pattern remains more or less stationary for a period in some region. Such a pattern is called a *block*, because the usual progression of weather systems from west to east is blocked off, or perhaps diverted farther north or south. The result is often that some areas, particularly those near centers of high surface pressure (anticyclones) experience clear skies and light winds for days on end. In the summer this means fine days, and in the winter frosty nights. However, in the same pattern, other areas are likely to experience prolonged gloomy conditions, or persistently strong winds, or periods of above-average rainfall.

One of the most important aspects of forecasting next week's weather is to predict any significant changes of type during the period. Especially after prolonged spells dominated by one particular weather type, customers place special value on receiving reliable advance notice that a change is on its way. Forecasters find it useful to have advice from two or even three different global models for this work, so medium-range forecasts are regularly exchanged between major forecasting centers. If the models agree – at least in general terms – then more confidence is placed in the forecast. When the models differ the forecaster has the difficult task of trying to combine their best features on the basis of previous experience of their strengths and weaknesses – a process requiring much skill.

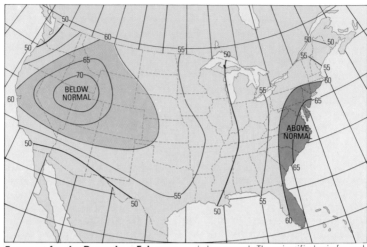

Forecast for the December–February season showed percentage probabilities that average temperature would be above or below normal. The scientific basis for such forecasts is fragile compared with the methods used for shorter-range forecasts.

The further outlook

Even with the fullest possible coverage of observations and the best possible global forecast model, there is a limit to the period for which a numerical forecast can be expected to be usefully accurate. With improvements gradually being made, the skill of forecast models is increasing toward this limit, believed to be about two weeks.

The limit to predictability is explained by the inevitable errors in the initial conditions. However small these errors are, they grow during the calculations and eventually they dominate the forecast. This suggests a technique that can provide additional information, though it is computationally expensive. Instead of running a single forecast for two weeks or a month ahead from a single set of initial gridpoint values, *several* forecasts can be run from slightly differing sets of initial values. For as long as these forecasts remain close to one another, reasonable confidence can be placed in their accuracy. But once they diverge significantly, as they eventually must, the limit of predictability has been reached. However, they continue to provide guidance on the *range* of the alternative weather conditions that are likely, and if the number of forecasts is large enough, probabilities could be assigned to the various outcomes.

On some days, the period of predictability will be longer than on others. Also, there are variations in predictability from one part of the world to another. Often the limit to predictability is shorter in the tropics than at higher latitudes, while certain weather patterns – particularly the blocking patterns mentioned on p. 133 – are predictable for longer than average.

An alternative approach to long-range forecasting is the attempt to exploit long-term records of past weather so as to develop effective methods of prediction. For example, at the end of a particular month it is possible to search through the records (a computer search is best of course) to find occasions when the same month in earlier years had similar weather. The suggestion is that the weather in the month just starting may have similarities with the weather in the corresponding month in the earlier years.

Many attempts have been made to forge useful statistical links out of the historical records so that sea surface temperatures, ice and snow cover, or various characteristic features of the large-scale patterns of atmospheric flow might be used as predictors of average conditions in the month or season which follows. As would be expected, the level of detail that is attempted in long-range forecasts is reduced quite considerably, even compared with that for forecasts a week ahead. Typically, a monthly forecast indicates, for broad areas, the probabilities of temperature and precipitation being above or below normal for the time of year.

It must be admitted that the success rates for long-range forecasts have been disappointing so far. Of the various predictors which have been investigated, probably the most promising is sea surface temperature. Volcanic eruptions may also play a part, injecting dust into the high atmosphere, where it intercepts the incoming solar radiation, but careful studies have failed to identify any significant effects on the subsequent weather.

Sea temperatures and long-range forecasting

In order to understand the behavior of the atmosphere over longer and longer periods, it becomes increasingly necessary to take the oceans into account. For studies of climate change over periods of years, fully coupled numerical models of the atmosphere and the oceans are certainly needed. For forecasting the general weather conditions a month or a season ahead, it may be possible to obtain useful guidance from an accurate knowledge of the global distribution of sea surface temperature at the start of the period.

Studies in several parts of the world have indicated connections between departures from the normal values for the time of year of sea surface temperature in particular regions and subsequent departures from normal of weather conditions elsewhere. Certain mid-latitude regions of the Pacific and Atlantic respectively are carefully monitored by long-range forecasters in North America and Europe for these "anomalies" of sea surface temperature. In tropical regions relationships have been found between sea temperatures and the tropical cyclones afflicting nearby land areas. If the sea temperatures before the start of a cyclone season are warmer than usual by as little as one degree, the number of cyclones coming on land during the season may be almost double the average.

Another relatively local impact of sea temperature anomalies in the tropics has been discovered through attempts to forecast the onset of the southwest monsoon over India. Once again, only one degree or so is involved, but warming the Arabian Sea by this amount can make all the difference to numerical forecasts of the onset as little as 1 to 2 weeks ahead.

The number of tropical cyclones observed each year in the Australian region, during the season October–May, increases with the anomaly from normal of the sea surface temperature north of Australia, averaged for the three months September–November.

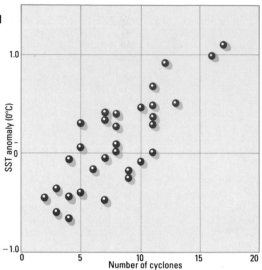

Sea surface temperatures in the tropical Atlantic and particularly in the tropical Pacific can have an important impact on the subsequent atmospheric flow patterns almost anywhere in the world. Research into these effects is code-named TOGA – Tropical Oceans, Global Atmosphere – and probably holds the key to several aspects of monthly and seasonal forecasting.

At irregular intervals of, typically, 3–7 years the tropical Pacific sea temperatures increase, in the East Pacific by perhaps 3 or 4 degrees, compared with normal. The effect is well known to the anchovy fishermen of Peru who christened it "El Niño" (Spanish: the Child) because the dramatic warming often reaches their fishing grounds during the Christmas season (the warming is associated with an almost complete disappearance of the anchovies). The El Niño phenomenon itself is a result of a large-scale, long-period interaction between the atmosphere and the ocean. Once the warmer water has appeared, however, the consequences for the global atmosphere one to three months ahead can be investigated by comparing long integrations of global forecast models with and without the observed sea temperature anomalies. The El Niños of 1973 and 1983 were associated with milder than usual winters over the United States, with more rain than usual in west coast regions. By contrast, the El Niño of 1977 was associated with an exceptionally severe winter over much of the North American continent. Studies using numerical models suggest that a crucial difference may have been that in 1973 and 1983 the warmer than normal water was confined to the tropical East Pacific.

The consequences of El Niños are not confined to North America. Indeed, one of the most important aspects of TOGA research is the possible link between anomalies in tropical sea temperatures and droughts in semi-arid regions such as the Sahel zone of Africa. Improvement in seasonal forecasting is quite literally a matter of life and death in such regions.

Supplying the customers

When members of the public sit down at home and watch TV representations of recent weather observations, sequences of satellite imagery and animated computer forecasts, it would be easy to forget the complex and expensive processes of information gathering and numerical prediction, and the applications of high technology, scientific expertise and forecasting skill that underlie what they see. The cost of all this would barely be justified if it only provided a public service, but fortunately the public service can be made available at little additional cost, once provision has been made for the important services for aviation, shipping, oil rigs, flood protection, industry, agriculture and so on.

Planning flights

Among the largest economic benefits of weather forecasting are those from services provided to aviation. Air safety is of course the primary consideration. Forecasts of cloud, visibility, and wind conditions at airports are required for the planning of take-off and landing schedules and diversions. Certain types of hazardous weather may be encountered at high altitudes during flights. These include clear air turbulence as well as the turbulence and

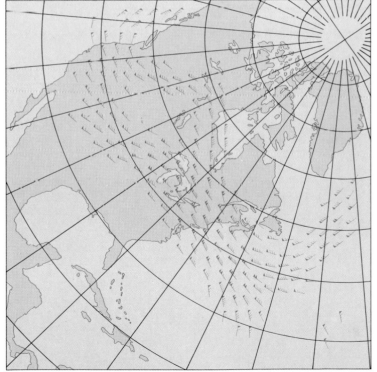

Forecast speeds and directions of winds over 60 knots are shown above by fleched arrows (triangles are 50 knots, full lines 10 knots and half lines 5 knots). In flight planning calculations, strong winds may either assist or oppose the airplane.

icing that occur within deep convective clouds that reach up to flight levels and beyond. Forecasts of the likely occurrences of such hazards are provided in the form of "significant weather charts" as part of the flight documentation for aircrew.

Gridpoint values of forecast winds and temperatures from global prediction models are passed across computer-to-computer links for use by aviation customers in flight planning calculations. The winds and temperatures encountered by an aircraft in flight lead to significant variations in the fuel required on a particular route from day to day. If an accurate weather forecast is available, the amount of fuel required can be calculated in advance. A suitable safety factor is allowed, of course, but the flight planning avoids the carrying of unnecessary fuel and thereby allows a bigger payload. Except in the busiest airlanes – where fixed tracks are laid down by air traffic control authorities – further savings can be made by choosing a route that avoids strong headwinds and takes advantage of any favorable tailwinds. Forecasts of jet streams are therefore particularly important for aviation. Strong crosswinds are also to be avoided where possible since, by requiring the aircraft to fly on a heading angled to the desired track (in order to stay on course), these also increase the consumption of fuel. In practice the flight planning software is used automatically to calculate the fuel consumption for a number of alternative routes, using the gridpoint wind and temperature forecasts, and the most favorable one is chosen.

Where fixed track flying is in operation, the air traffic control authorities also make use of the meteorological forecasts to ensure that the required separations between aircraft are maintained. Accurate forecasts allow a denser flow of air traffic, without increasing the risk to safety, and so help to avoid delays.

Shipping forecasts

It was concern for the safety of ships that prompted the first developments of meteorological services in the middle of the nineteenth century, and the shipping and fishing industries remain important customers today. Winds, waves, and visibility are specially important, and in most parts of the world there are routine radio broadcasts that summarize the expected conditions over the next day or so in a series of internationally agreed sea areas. These forecasts are of great value in assisting mariners to avoid gales and other severe conditions if possible, or otherwise to take the necessary precautions to reduce damage to cargo and danger to the vessel. At high latitudes in winter, warnings of the danger of icing are included in shipping forecasts.

A more recent addition to forecasting services for shipping is weather routing. Forecasts of the winds and waves expected over the major oceans are used to provide advice on the best route to take. For example, for voyages to and from many ports in northern Europe there is a choice between a route through the English Channel and an alternative route passing to the north of Scotland. The difference in the distance to be travelled may be slight, but if one route involves a passage through stronger winds and heavier

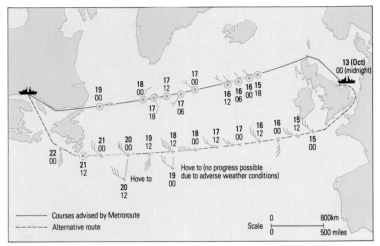

The ship that followed the advised route (north of Scotland) gained about 2 days on the voyage compared with the alternative route (English Channel) along which strong opposing winds and waves would have been encountered.

seas than the other, significant delays may be avoided by the weather routing service. An additional important factor is the damage to cargoes that may occur in heavy seas.

In other cases there is no such choice of alternative routes resulting from the geography, but weather routing of ships still has an important role. For example, in Pacific crossings from the west coast of North America to Japan or Southeast Asia the preferred route in calm conditions would obviously be the shortest, known as the great circle route. But the expected weather conditions and sea state along the shortest route may be so disadvantageous that time can be saved by taking a longer route that avoids strong winds and heavy seas.

For these very long sea crossings, lasting perhaps for 10 or more days, there is the problem that the weather forecast for the end of the period is inevitably less accurate, as discussed on pp. 121–122. The ships' routing service therefore includes regular updates during the course of the voyage, and the route originally planned is modified if the later forecasts show this to be advantageous.

Quite frequently there are cases where study after the event shows that two or three days were saved on an Atlantic or Pacific crossing by following the advice of the weather routing service. This is obviously of major economic importance to shipping companies.

The quality of shipping forecasts depends mainly on the accuracy of numerical forecasts of the surface wind over the sea. Numerical models are often designed with several levels close to the surface, and this contributes to improved accuracy. Both global models, for medium-range forecasts over the world's oceans, and limited-area fine-mesh models, for coastal aspects, are important for services to shipping.

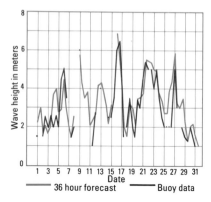

Off-shore oil rigs' operation and supply require forecasts of weather and wave conditions. Results from numerical wave prediction models are compared with measurements from instrumented buoys (above) to demonstrate their usefulness in support of such services.

Warning oil rigs

The rapid expansion of the offshore oil industry in several parts of the world during the 1970s made fresh demands upon meteoroogical services. Certain phases of offshore work are extremely sensitive to weather and wave conditions, including the initial surveys, the exploratory drillings, the towing out and erection of structures, and the full production phase. Unplanned interruptions to the work are very costly, and there is risk of serious damage to equipment and even loss of life if severe conditions occur without warning.

Winds are obviously of key importance, as is the wind-sea, that is the waves generated locally by the winds. But possibly the most serious hazard comes from large waves that arrive when the local winds are light, having been generated by a storm hundreds or even thousands of miles away. Offshore structures are often particularly sensitive to these swell waves, which are likely to have longer wave lengths and lower frequencies than most of the locally generated waves.

The difficult problem of swell prediction was largely overcome by the development of numerical wave prediction models. Using surface wind forecasts from atmospheric models as input, the wave models calculate the changes at each gridpoint of the energy contained in waves at a number of different frequencies and at regular

intervals of direction. In this way swell waves, generated by very strong surface winds in depression or hurricanes, can travel huge distances along great circle tracks in the numerical calculations.

Regular forecasts of wind, wind-sea, and swell at frequent intervals are transmitted to the oil rigs as routine. Verifications using measurements from specially instrumented buoys show that the wind and wave forecasts from numerical models run on powerful supercomputers are superior to anything that could be achieved by human forecasters working without such calculations. Once again, however, the forecaster has a valuable role in assessing and interpreting the numerical guidance. For specially sensitive operations such as towing out a large structure, for which a "weather window" of calm conditions over several days may be required, the best practice is for a forecaster to be attached to the operations team for the duration of the task to provide the latest advice. Thus his contribution would be tailored directly to the special needs of the work.

The wave forecasting models find a rather different application when they are run for long periods using past wind data. These *hindcasts* are used along with the small quantities of measured wave data that are available to establish climatologies of wave conditions for design and certification of oil rigs and other off-shore structures.

Watching for floods

Certain populated regions of the world are at risk from flooding under particular weather conditions. Clearly it is important that warnings be issued as early as possible, particularly to the authorities responsible for initiating precautionary measures against injury to persons and damage to property.

There are three kinds of flooding: flooding by rainfall, flooding by snow-melt, and flooding by the sea. Heavy rainfall may lead directly to flash floods in extreme cases (such as intense thunderstorms) when the surface water cannot run off quickly enough through its normal channels. More frequently, flooding of streams and rivers occurs after a prolonged period of heavy rainfall. To provide a basis for warnings of these floods, the requirement is for accurate rainfall forecasts. The main tools at the forecaster's disposal here are fine-mesh numerical models (see p. 119), radar measurements, and the special processing of geostationary satellite measurements. In regions where flooding by snow-melt is a hazard, accurate temperature forecasts are the primary meteorological requirement.

Coastal flooding has two main aspects. First, the water level at the coast may increase beyond its expected tidal range as a result of the meteorological conditions. When there is a difference between the water level and the astronomically predicted tide, this

The construction of the Thames barrier (right) indicates the serious threat of tidal flooding in central London. The barrier is operated on the basis of warnings issued by forecasters who use computer models of tidal flow to calculate the storm surges due to forecast winds and pressures.

is known as a storm surge. If the surge is big enough and coincides closely enough with high tide, the water level may exceed the coastal defenses that have been provided, with possibly disastrous consequences. The second aspect of coastal flooding is the risk that waves superimposed on the general water level may overtop the coastal defenses. In such cases it is the swell wave (see p. 140) that poses the greatest danger, for such waves may arrive at the coast quite unexpectedly, when the local wind is light and conditions are apparently harmless.

To provide a warning service for coastal flooding, forecasters make use of two kinds of oceanographic prediction model, both driven by inputs from atmospheric models. We have already mentioned wave models, driven by surface wind forecasts, in connection with services to oil rigs. The same models can provide information relevant to coastal flooding. Storm surge models require data on surface winds and also surface pressures from atmospheric models; they use these data to calculate estimates of the water level at each gridpoint. Here again it is the quality of the surface wind forecasts that mainly determine the quality of the service provided. At times of danger, specialist forecasters keep constant watch on telemetering tide gauges, capable of directly transmitting data, to assess the quality of the numerical forecasts and issue amendments to warnings as necessary.

Road transport can be disrupted by severe winter weather, affecting many sectors of industry and commerce. Specialized weather forecasts can be used to alert the relevant authorities that precautions may be necessary to keep more roads open more often.

Control rooms of the electricity generating industry use weather forecasts of temperature in particular, in combination with other factors, to estimate demand in the hours and days ahead. Production levels are adjusted accordingly.

Advising industry and commerce

Many aspects of industry and commerce are sensitive to the weather. Meteorological observations are of value in providing several types of advisory service in these areas. The design of structures, the siting of weather-dependent industries, and estimation of demand for products and services all need climatological data from long series of weather observations. Detailed records of the weather at particular locations during specified periods, corresponding perhaps to a particular construction project, are valuable when it comes to assessing the likely effect of adverse weather as regards delay in the completion of outdoor work. Knowledge about weather conditions at specific times and places is often required in connection with insurance claims and legal disputes.

There is also major economic value in forecast weather information. In many cases, notably that of the construction industry, the forecasts must be specific to the site of operations and must highlight the weather elements – wind, temperature, rainfall and so on – to which the work is most vulnerable. Such forecasts are provided by specialist meteorologists who interpret the computer forecasts along with radar and satellite imagery.

Severe weather can seriously affect the transport aspects of almost all industries. To avoid disruption of road and rail transport by ice and snow, those responsible must have access to area forecasts. Special scientific problems may also be involved, such as the prediction of road surface temperatures.

The prediction of demand is the objective in an important group of applications of weather forecasts to industry and

commerce. The gas and electricity industries, for example, have long made use of temperature forecasts in the calculations of energy demands over periods of a few days. More recently, similar techniques have been introduced in the food retail and distribution industries, permitting managers to make informed choices between alternative distribution schedules according to the expected demand for, say, ice cream.

In the financial world, futures markets in several commodities are also now influenced by weather forecasts, and services have been provided to cater to these requirements: coffee is affected by forecasts of frost in Brazil, and oil is affected by forecasts of low temperatures in New York.

As more attention is paid to renewable sources of energy, weather information is becoming even more important. Wind energy conversion systems (WECS) are already making a useful contribution to the total energy requirements of several countries. In this case the weather not only affects the demand for energy; it also controls the production capability. Thus for operating an integrated energy supply system, temperature forecasts are required for major population centers, and wind forecasts (at a level of about 50m (160ft) above the ground) at WECS sites need to be regularly provided.

Wind turbines like these operating in California generate electricity without the problems of fossil fuels and nuclear power, and are likely to become increasingly important. Meteorological records over many years are used when choosing sites for such ''wind farms''. Then, when the turbines are being operated in an integrated system that uses energy from various sources, wind forecasts are needed for predictions of production rates.

Helping the farmers

In all parts of the world, farmers are dependent on the weather for the success of their work. At the same time the weather often poses severe problems which the farmer must struggle to overcome.

The introduction in recent years of useful weather forecasts extending to about a week ahead has been a major advance in terms of meteorological services to agriculture. Of course, monthly and seasonal forecasts would be even more significant, but as we have said (see pp. 135–6) these will have to wait until further research bears fruit. But even a week ahead, much can be done at critical periods, such as during a crop harvest, to avoid damage to crops and wasted work if guidance on periods of wet and dry weather is reliable. Warnings of likely wind or hail damage can also enormously help harvesting plans.

Almost every aspect of the weather can be important to the farmers at various seasons of the year, affecting a wide range of crops and livestock. Frost warnings, for example, are vitally important to many growers of fruit and vegetables, particularly when expensive precautions must be taken – as with the oranges of Florida. For animal farmers the combination of strong winds and low temperatures spells danger to vulnerable livestock; here the weather warnings are often expressed in terms of wind chill factors

Damage from hailstones is perhaps the most dramatic effect of weather on crops. Hail warnings can have a big impact on harvesting decisions.

which have been determined through special research for various species.

Irrigation is another expensive process, but one which may be amply repaid in terms of increased crop yields. Meteorological advice on when to irrigate, and how much water to use, is based partly on calculations of the soil moisture conditions that result from the accumulated effects of rain, sun and wind over many months, and partly on forecasts of rainfall amounts during the coming few days.

Large-scale spraying of crops, using light aircraft to apply pesticides or nutrients, can only be carried out in suitable weather conditions. Here the wind speed is all-important; if it is not below a certain threshold value the spray will be dispersed away from the crop.

Wind forecasts are also important for predicting the likely spread of airborne diseases from one place to another. Low level air trajectories can conveniently be calculated from numerical forecast data for this purpose. Other meteorological elements such as temperature, humidity and rainfall are used in the preparation of warnings of diseases and pests that can only multiply if sufficiently favorable conditions occur. For example, desert locusts begin to breed about the time of the onset of rain in previously dry areas. Thus, if rainfall in breeding areas is accurately predicted, or

The light aircraft that are used for the
spraying of crops need local forecasts of
cloud, visibility and wind for safe and
effective operations.

at least rapidly detected from satellite images, action can be taken
to prevent swarms developing.

Telling the public

Last, but not necessarily least, weather information is made available to the general public in every country in the world, and free
access to weather forecasts is widely regarded as a citizen's right.
Sometimes the weather forecast can be of great importance to
members of the general public, particularly when it involves warnings of gales, frost, fog, blizzards, thunderstorms and other
weather hazards. In the case of hurricanes, tornadoes and floods
the weather forecasts may be a matter of life and death. Forecasts
of sun and snow are very important in relation to pleasure resorts
too. Nevertheless, for most members of the general public most of
the time the weather is unlikely to amount to more than an inconvenience or perhaps a minor discomfort. But it is a fact that there is
often a considerable social interest in the weather forecast – and
particularly in any errors in the forecast!

The methods available for telling the public about the weather
are familiar and straightforward – newspapers, radio, television,
telephone and teletext. Each has its advantages and disadvantages. Newspapers and television reach large sections of the
public, and can use maps to present the forecast clearly and in

TV weather presenters have become familiar figures in homes all over the world, as they retail in simple terms the implications of complex calculations.

detail; but newspapers can only update information once a day, whilst on television the time available for presentations is often very restricted – hence the development of specialist weather channels in the USA. On radio, time is often more generous, and radio probably reaches the greatest numbers of people at many times of the day or night, including many traveling by road. However, here there is sometimes difficulty in communicating information clearly unless the region of coverage is highly local. Recorded telephone messages are very effective for detailed forecasts in a local area. In principle, a teletext or viewdata system can combine the best features of all the other methods, being frequently updated and giving, on different pages, specific information for particular areas and for particular activities and recreations. But teletext reaches relatively few people as yet.

Among other benefits, modern methods of presenting weather forecasts to the public have had an educational value. Regular displays of satellite imagery, for example, have succeeded in conveying an impression of the frequently patchy nature of cloud cover which is such a headache for forecasting. Similarly, animations based on computer forecasts emphasize the all-important movement of weather patterns – a factor that had not been widely appreciated before.

Generally speaking, the public seems to be impressed at the increasing accuracy of forecasts that are presented. But the weather still keeps its surprises. Forecasts will improve still further, and the surprises will get fewer as a result. But the fascinations of the weather, how it works and how it changes, seems unlikely to diminish.

GLOSSARY

Accretion The growth of a frozen particle (i.e. ice crystal or snowflake) by its collision with a **supercooled** liquid droplet that freezes on contact.

Acid rain Rainfall with a high acidity, probably caused by industrial effluents.

Adiabatic cooling In a system where there is no transfer of heat across its boundaries, changes of pressure cause changes of temperature. Compression causes warming, expansion causes cooling. Hence air that rises expands and cools.

Advection The horizontal transport of an atmospheric property by the air flow.

Aggregation When ice crystals meet and stick together they are said to "aggregate".

Air mass A large volume of air within which the horizontal gradients of temperature and humidity are relatively small. Tends to form over fairly homogenous surfaces such as ice caps and oceans.

Albedo The ratio of the amount of radiation reflected by a body to the amount received by it – usually expressed as a percentage.

Anabatic Upslope.

Anticyclone A weather system with relatively high atmospheric pressure (often called a "high"). Anticyclonic winds blow clockwise in the northern hemisphere and counterclockwise in the southern hemisphere. Air settles downwards and diverges outwards with a cloud-dissipating effect.

Arctic smoke Fog formed when water vapor is added to air which is much colder than the water vapor's source.

Boundary layer In the atmosphere, the bottom 3,000ft (1km) of air, the boundary being the Earth's surface.

Buoyancy force The upward force exerted upon a parcel of air by virtue of the density difference between the parcel and that of the surrounding air. If the air in a balloon is lighter (less dense) than its surrounds, the balloon will rise; if the air is heavier (more dense), the balloon will sink.

Coalescence In cloud physics the merging of two water droplets to create a larger one.

Condensation The process by which vapor becomes a liquid or a solid.

Conduction The transfer of heat by means of the movement of internal particles and without any net external motion. It is the mechanism whereby, for example, the handle of a poker in a fire becomes hot.

Convection The transfer of heat by motions within the air, predominantly vertically, to distinguish it from **advection.**

Convergence The contraction of the field of flow, such as occurs when water flows down the drain of a household sink.

Coriolis force An effect due to the Earth's rotation that affects the direction of horizontal airflow. In the northern hemisphere it "deflects" air to the right and in the southern hemisphere it "deflects" it to the left.

Coriolis scale A scale of atmospheric flow at which the coriolis force plays a major role. Same as "synoptic scale".

Cyclone A weather system consisting of an area of low pressure, often known as a "low". Winds blow counterclockwise in the northern hemisphere and clockwise in the southern hemisphere. Same as a depression.

Deformation The change in the shape of a mass of air due to spatial variations in its velocity, particularly stretching or shearing. Thus, if a cream cake is sat on, the cream filling is squashed out sideways, and the whole cake is deformed.

Density Mass per unit volume.

Depression An area of relatively low pressure, usually associated with cyclonic circulation.

Dew-point The temperature to which a parcel of air with a constant water vapor content must be cooled at a constant pressure in order for saturation to occur.

Divergence The spreading of a velocity field and thus the opposite of **convergence.** If applied to a mass of air, it becomes a measure of the rate of net transfer of mass out of a specified volume of space.

Doldrums A nautical term for the low pressure area along the equator which has light winds.

Downslope wind Any wind which moves downslope, as in mountain areas such as the Alps (Föhn) or the Rockies (Chinook).

Dropsonde A device dropped from research aircraft to measure temperature, pressure and humidity values (and sometimes wind speeds) at various levels in the air below. The values are radioed back to the aircraft or to a ground station.

Evaporation The process whereby a liquid or solid is transformed to a gas.

Fine-mesh model A type of numerical forecasting model in which the spacing of the **gridpoints** is reduced to achieve greater accuracy and detail. Because of the increased computing costs, the coverage is limited to a region of special interest.

Föhn wind (Chinook) A warm, downslope wind resulting from **adiabatic** warming on its downward path. Found north of the European Alps and in Alberta among many other places to the lee of mountains.

Front The transition zone between two air masses of different temperatures and hence density.

Geostrophic wind The horizontal wind resulting from a balance between the pressure gradient and coriolis forces.

Greenhouse effect The heating effect exerted by the atmosphere upon the Earth due to the fact that the atmosphere absorbs and re-emits infrared radiation. This re-emitted radiation helps to keep the Earth's temperature some 40°C higher than it would be under simple radiative equilibrium. The effect is so-called because the atmosphere acts in the same way as do panes of glass in a greenhouse.

Gridpoint A geographical location at which the values of variables are calculated in a numerical forecasting model.

Hindcast A calculation, made after the event, of the sequence of values of a variable that has not been measured directly. For example, the calculation of sea-state from observed values of the surface wind.

Humidity A measure of the water vapor in air.

Infrared Applied to wavelengths of radiation. Infrared describes the wavelengths from 0.8 micrometres to an indefinite upper limit of about 100 micrometres.

Inversion A departure from the usual decrease or increase with altitude of the value of an atmospheric property. A temperature inversion means an increase with height.

Isobar A line joining equal values of pressure.

Jet stream Relatively strong winds concentrated within a narrow stream in the atmosphere.

Katabatic Moving downslope.

Land breeze A land-to-sea nocturnal breeze.

Latent heat The heat released or absorbed when water is condensed or evaporated.

Lee wave A wave in the airflow to the lee of the obstacle disturbing the flow.

Mechanical Effects of energy and forces upon moving bodies. Often used to differentiate from thermally-induced effects in the atmosphere.

Meridian Line of longitude.

Meridional flux Transfer of an atmospheric property along the meridians.

Mesoscale A description of horizontal dimensions of some atmospheric systems. Refers to systems 100km or so across.

Microscale A description of horizontal dimensions of some atmospheric systems. Refers to systems a few meters, or less, across.

Millibar A unit of atmospheric pressure.

Monsoon A seasonal wind.

Net radiation The difference between incoming and outgoing radiation.

Nowcast An estimate of what is happening now, or a very short time ago, based on whatever observations are available in real time and on diagnostic relationships between these observations and other variables of interest. One of the best known examples is the estimation of rainfall intensities in a remote area from radar and satellite imagery.

Occlusion A composite of two fronts formed as a cold front overtakes a warm front in an extratropical cyclone.

Orography The varied elevation of terrain.

Pressure Force per unit area. In the atmosphere the mass exerts a force and this is measured in terms of pressure.

Pressure gradient The change of pressure over a specified distance, usually horizontal or vertical.

Radiation Transfer of heat by electromagnetic waves. These waves permit the propagation of energy through space, in contrast to **convection** and **conduction** which require a medium through which heat is transferred.

Radiometer An instrument to measure the intensity of **radiation** of a particular type (e.g. solar, infrared). It may be ground-based or carried on an aircraft or a satellite.

Radiosonde A package of instruments carried aloft by balloon to measure temperature, pressure and humidity values (and sometimes wind speeds) at various levels in the air above. The values are radioed back to a ground station, possibly via a satellite link.

Ridge An elongated area of relatively high pressure, sometimes known as a wedge.

Satellite (geostationary) A satellite which always stays above the same point on the Earth's surface. This is possible only for points on the equator, and only if the satellite is in orbit at 22,500 miles (36,000km) above the Earth.

Satellite (polar-orbiting) A satellite whose orbit takes it over both the Arctic and Antarctic regions in turn whilst the Earth rotates below.

Saturation In the atmosphere this term almost always applies to water vapor. It is a state such that the pressure exerted by the vapor is equal to its maximum under any particular condition, such that any increase would lead to **condensation** of water into liquid.

Saturation vapor pressure	Almost always applied to water vapor in an atmospheric context. It is the pressure, at a given temperature, exerted by the vapor when it is saturated.
Sea breeze	A daytime breeze blowing from sea (or lake) towards land.
Sensible heat	The heat of air that we feel and that is measured by the dry-bulb thermometer. Contrasts with latent heat.
Smog	A combination of smoke and fog.
Supercomputer	A computer which has been specially designed and constructed to carry out large numbers of arithmetic calculations on large volumes of data very quickly.
Supercooled	The state of a substance (water in this context) cooled to below its freezing point yet remaining as a liquid.
Supersaturated	The condition existing in a given portion of the atmosphere when it contains more water vapor than is needed to produce saturation with respect to a flat surface of pure water or ice.
Synoptic	In meteorology, refers to the use of data obtained simultaneously over a wide area.
Temperature gradient	The change of temperature over a specified distance.
Thermal	A small volume of rising air produced when the atmosphere is unstable. One of the elements of the convection process.
Time-step	An interval of time (typically in the range from a few minutes up to an hour) used when stepping forward the values of the variables in a numerical forecasting model.
Trade winds	The wind system occupying most of the tropics blowing from the subtropic highs into the Inter-tropical Convergence Zone. The winds are northeasterly in the northern hemisphere and southeasterly in the southern hemisphere.
Translation	Motion of a body in a straight line, such as followed by a bullet fired from a gun.
Transpiration	The process whereby water in plants is transferred as water vapor to the atmosphere.
Troposphere	The lowest 6 to 12 miles (10 to 20km) of the atmosphere; shallower over the poles; deeper over the equator. This layer contains the weather as we know it.
Trough	An elongated area of relatively low pressure.
Vorticity	A measure of rotation in a fluid flow. By convention, cyclonic vorticity is positive and anticyclonic vorticity is negative.

Index

W

warm fronts, 13, 27
cloud formation, 78
warm sectors, 27
water circulation, 68–94
cloud classification,
81–7
cloud formation,
77–81
dew, 72
evaporation, 68–9, *68*
fog, 72, *72–6*, 75–7
humidity, 70–1
precipitation, 88–94
smog, 72
transpiration, 68, *68*,
69
wave forecasting, 140–1
weather charts, 12–13
weather routing, 138–9,
139
weather stations, ocean,
103–5, *104*
Weather Wire, 16

wet-bulb thermometers,
70
wind energy conversion
systems (WEVS), 146,
147
wind speed:
measurement, 98–9
satellite measurement,
107
upper air
measurement, 100,
105
wind turbines, *147*
winds, 58–66
anabatic flow, 33
doldrums, 60
downslope, 30, 32, *32*
extremes of, 64–6, *65*
Föhns, 32, 71
friction, 60
geostrophic, 59, *59*
global pattern, 56
gradient, 59
hurricanes, 24
jet streams, 60, 61, *61*

katabatic flow, 34
land breezes, 33, *34*,
61
monsoons, 44
patterns, 9–10
pressure and direction
of, 12–13, 59
pressure gradient,
58–9
prevailing polars, 60
prevailing westerlies,
60
sea breezes, 33, *34*, 61
thermal wind
mechanism,61
trade, 9, *41*, 43, 56, 60
wooded areas,
microclimates, 38
World Area Forecast
Center (WAFC), 124
World Climate Research
Program, 117
World Meteorological
Organization (WMO),
16, 98, 125

Acknowledgments

T = top   **B** = bottom   **C** = center   **L** = left   **R** = right
Cover: Tony Stone Photo Library. **Title Page:** Maurice Nimmo/Frank Lane
Picture Agency. **Contents Page:** Vautier-De Nanxe. **14–15** ZEFA; **18–19** NOAA;
20–1 European Space Agency; **22–3** ZEFA; **24–5** ZEFA; **30** ZEFA; **31**Nicholas
Devore/Bruce Coleman; **35** Wil Blanche/Rex Features; **36–7** H. Binz/Frank Lane
Picture Agency; **37** Maurice Nimmo/Meteorological Office; **39T** ZEFA; **39B**
Bernard Regent/The Hutchison Library; **40–1** ZEFA; **43** TRH Pictures; **45**
Vautier-De Nanxe; **46** US Naval Research Laboratory; **50–1** Spectrum Colour
Library; **52–3** Bryan & Cherry Alexander; **53** Tom Nebbia/Aspect Picture
Library; **54** ZEFA; **56** Bryan and Cherry Alexander; **57** S. D. Burt/Meteorological
Office; **58–9** Spectrum Colour Library; **60–1** NASA; **63** Crown
Copyright/Meteorological Office/By permission of Controller of HMSO; **65T**
Herman Kokojan/Black Star/Colorific; **65B** Frederick Ayer/Science Photo
Library; **66–7** Art Directors Photo Library; **67** P. J. May/Meteorological Office;
70–1 Armstrong/Zefa; **72** ZEFA; **73** Manheim/ZEFA; **74** J. Alex Langley/Aspect
Picture Library; **75** Crown Copyright/Meteorological Office/By permission of
Controller of HMSO; **76T** A. Honeyman/Meteorological Office; **76B** Art
Directors Photo Library; **77** T. A. M. Bradbury/Meteorological Office; **78** S. D.
Burt/Meteorological Office; **79** Steve McCutcheon/Frank Lane Picture Agency;
80 NASA; **81** M. P. Price/Bruce Coleman; **82–3** Spectrum Colour Library; **83T**
Jonathan Wright/Bruce Coleman; **84** NASA/Science Photo Library; **84–5** N.
Elkins/Meteorological Office; **85T** Tony Morrison/South American Pictures; **86–7**
John G. Ross/Susan Griggs Agency; **86** Kalt/ZEFA; **90** Nicholas Devore/Bruce
Coleman; **92** Lee Lyon/Bruce Coleman; **93** Erwin & Peggy Bower/Bruce
Coleman; **95** Popperfoto; **96–7** Crown Copyright/Meteorological Office/By
permission of the Controller of HMSO; **99** R. Halin/ZEFA; **101** Crown
Copyright/By permission of Controller HMSO; **110TL** Royal Meteorological
Society; **110TR** Royal Meteorological Society; **111** Crown Copyright/By
permission of Controller of HMSO; **113** NASA; **115** Meteorological Office;
116TRH Pictures; **117** M. J. Kerley/Meteorological Office; **128–9** Meteorological
Office; **131** Crown Copyright/Meteorological Office/By permission of Controller of
HMSO; **140–1** Shell Photo Service; **142–3** Handford Photography; **144**
Popperfoto; **145** CECG; **146–7** ZEFA; **148** J. C. Allen & Son/Frank Lane; **149**
Holt Studio Ltd; **150** CBS News.